PHILOSOPHICAL INTRODUCTION TO SET THEORY

Stephen Pollard
Truman State University
Kirksville, Missouri

Dover Publications, Inc.
Mineola, New York

Copyright

Copyright © 1990 by Stephen Pollard
All rights reserved.

Bibliographical Note

This Dover edition, first published in 2015, is an unabridged republication of the work originally published in 1990 by the University of Notre Dame Press, Notre Dame, Indiana.

Library of Congress Cataloging-in-Publication Data

Pollard, Stephen (Stephen Randall)
 Philosophical introduction to set theory / Stephen Pollard. — Dover edition.
   pages cm
 Originally published: Notre Dame, Indiana : University of Notre Dame Press, 1990.
 Includes bibliographical references and index.
 ISBN-13: 978-0-486-79714-4
 ISBN-10: 0-486-79714-7
  1. Set theory. I. Title.
 QA248.P657 2015
 511.3'22—dc23

2015000846

Manufactured in the United States by Courier Corporation
79714701   2015
www.doverpublications.com

To Norman M. Martin

CONTENTS

Preface — xi

I. Queen of the Formal Sciences — 1
   1. Mathematical Glue, 1
   2. Mathematics' One Foundation, 4
   3. Science of the Infinite, 6
   4. But What Are Sets?, 12

II. Offspring of Analysis — 13
   1. A Role for History, 13
   2. Greek Roots, 15
   3. The Latitude of Forms, 17
   4. Eidetic Logistic, 20
   5. *Genitum* and *Functio*, 23
   6. Functions as Notation, 25
   7. Cantor the Analyst, 27
   8. Numbers, Functions, Sets, 32
   9. Dedekind Cuts, 33
   10. Frege on Abstraction, 36

III. Commonsense Sets — 40
   1. Nausea, 40
   2. The Court of Ordinary Language, 41
   3. Black on Sets as Commonsense Objects, 42
   4. Sets and Plural Reference, 44
   5. Zermelian Sets, 51

viii CONTENTS

IV. LOGICAL INTERLUDE: INTERPRETABILITY   56

    1. The Age of Logic?, 56
    2. Interpretation Functions, 57
    3. Abstraction as Interpretation, 59
    4. Pseudo-Uniqueness, 63

V. FORMALISM   65

    1. Universal Theories as Mathematical Objects, 65
    2. Cantor's Continuum Hypothesis, 67
    3. The Robinson-Cohen Challenge, 70
    4. A Dummettian Argument, 72
    5. Why Bother?, 79

VI. SOME METAMATHEMATICS   82

    1. A Formal Language and Its Logic, 82
    2. The Principle of Comprehension, 85
    3. Interpretation of the Successor Function, 89

VII. LOGICAL INTERLUDE: SECOND ORDER LOGIC   97

    1. Logics of the First and Higher Orders, 97
    2. A Monadic Second Order Deductive System, 100
    3. Boolos on Monadic Second Order Logic, 103
    4. Second Order Z, 106
    5. CH Is Second Order Decidable, 110

VIII. ITERATIVE HIERARCHIES   115

    1. Set Formation, 115
    2. Well-Ordered Structures, 117
    3. Choosing a Theory, 119
    4. Theory T_1, 121
    5. Theory T_2, 129
    6. A Modification of T_2, 133
    7. Universal Structures, 135

IX. STRUCTURALISM   138

    1. Stenius on Sets and Structures, 138
    2. Reference and Reduction, 143
    3. Resnik's Mathematical Structuralism, 148

APPENDIX I   155

APPENDIX II   161

NOTES   165

BIBLIOGRAPHY   173

INDEX OF NAMES   179

PREFACE

A. J. Ayer has described his *Language, Truth and Logic* as "in every sense a young man's book" – by which he means that it is passionate and dogmatic to the point of being offensive. This is indeed one form which a "young man's book" is likely to take. But even a fledgling philosopher can appreciate the maxim: To love wisdom is not yet to be wise. By keeping this dictum constantly in mind, I have perhaps been led to produce a book which is annoying for its *lack* of dogmatism. For my principal aim has not been to propound universal law, but to remove impediments to rational discourse. Philosophers interested in set theory currently form a scattered choir of dissonant and confusing voices. By offering a clear statement of basic questions and problems, I hope to make the philosophy of set theory somewhat less anarchic and somewhat more fruitful. I am convinced that this is a necessary condition for an adequate philosophical account not just of set theory, but of mathematics in general.

Set theory is not merely a specialized mathematical field. It is the primary mechanism for ideological and theoretical unification in modern mathematics. Any adequate treatment of the philosophy of mathematics will have to feature a technically informed discussion of set theory. And no technical treatment of set theory will be entirely satisfactory if it ignores the wide-ranging philosophical issues surrounding set theory's foundational status. It is therefore crucial that students have available a book-length treatment of set theory which is intelligible to educated non-specialists and which deals with philosophical issues in a technically sophisticated way. This volume is meant to meet precisely this need. It is intended to be accessible to graduate students and advanced undergraduates who have some aptitude at mathematical reasoning and some prior exposure to symbolic logic. It should thus be suitable as a source of supplementary readings in a course on set theory or as a central text in a course on the philosophy of mathematics.

Aesthetic considerations alone have spared this book the more accurate, but less euphonious title "Philosophical Introduction to Neo-Classical, Impredicative Set Theory." So let the reader now be given fair warning: In what follows, I shall largely ignore intuitionist and constructivist alternatives to the Cantorian-Zermelian tradition. Attempts to insinuate modal logic into this tradition will be similarly slighted – the charms of the set theoretic mainstream having diverted me from any intensive exploration of the tributaries.

Chapter 8 of this volume features excerpts from my "Plural Quantification and the Iterative Concept of Set" (*Philosophy Research Archives* 11 [1986]: 579-587; © 1986 by the Philosophy Documentation Center). I am grateful to the Philosophy Documentation Center for permission to use this material. A slightly abridged and altered version of my "Plural Quantification and the Axiom of Choice" (*Philosophical Studies* 54 [1988]: 393-397; © 1988 by Kluwer Academic Publishers) appears in §4 of chapter 7. This material is reprinted here by kind permission of Kluwer Academic Publishers. Finally, several paragraphs from "Mathematical Naturalism: An Anthropological Perspective" by Robert Bates Graber and myself (*The Southern Journal of Philosophy* 27 [1989]: 427-441) appear in chapter 2 by permission of the editor of the *Southern Journal of Philosophy*.

I owe thanks to Patricia Burton and Daniel Bonevac – who read most of an early version of this work and supplied incisive comments. The perceptive remarks of several anonymous referees also contributed significantly to this book's current form. My own introduction to set theory (and a most philosophical introduction it was) was supplied by Norman M. Martin – to whom this volume is dedicated.

I
QUEEN OF THE FORMAL SCIENCES

1. Mathematical Glue

In contemporary mathematics, set theory reigns supreme. Testimonials to this effect can be cited *ad nauseam:*

> As one of the characteristic features of modern mathematics, we must ... mention the obvious dominance of the set-theoretical point of view.[1]

> Set theory is the foundation of mathematics. All mathematical concepts are defined in terms of the primitive notions of set and membership.[2]

> The idea of a set is basic to all of mathematics, and all mathematical objects and constructions ultimately go back to set theory.[3]

> By analyzing mathematical arguments, logicians became convinced that the notion of "set" is the most fundamental concept of mathematics.[4]

> The ideas and concepts of the theory of sets penetrated literally into all branches of mathematics and changed its face entirely. Therefore it is impossible to form a proper picture of contemporary mathematics without being acquainted with the elements of the theory of sets.[5]

Many mathematicians manage to avoid spontaneous declarations of allegiance to Her Majesty. They instead fill their lectures and papers with topics and terminology which bear no obvious relation to set theory. But were you to ask such mathematicians about the ultimate foundations of their work, you would probably get an earful of set talk on the spot – or, at the very least, would be sent scurrying down the hall to confer with the nearest set theorist. In one way or another, the knee is bent and the proper respects are paid.

What is this wonderful thing called "set theory"? And what does it *do* to merit such homage? Set theory, first of all, is a powerful glue helping to keep mathematics together as a single science in the face of powerful tendencies toward disintegration. Chief among these tendencies is a severe narrowing of individual areas of competence which threatens to break the mathematical community into intellectually isolated groups of specialists.

A bit of history may help to drive this point home. In 1868, there were about 38 recognized areas of specialization within mathematics. In 1979, there were roughly 3,400.[6] This proliferation of subfields within areas which were themselves already fairly narrow can occur with disconcerting speed. In the early 1960s, a bright, well-read logician could, with a modest expenditure of time and mental energy, thoroughly understand any article published in the *Journal of Symbolic Logic*. By the mid-1970s, a merely superficial grasp of a *JSL* paper had become a pleasant alternative to one's all too frequent bouts of utter incomprehension. (A related statistic: The 1946 volume of *JSL* contained 9 articles spread over 144 pages. The 1986 volume offered 91 articles and 1,145 pages.) As one prominent mathematician notes (with what seems to be a mixture of admiration and revulsion), "The variety of objects worked on by young scientists is growing exponentially. Perhaps one should not call it a pollution of thought; it is possibly a mirror of the prodigality of nature which produces a million species of different insects. Somehow one feels, though, that it goes against the grain of one's ideals of science."[7]

The point here is not simply that more and more mathematics is being produced, but rather that the number of people in a position to understand any given result is growing smaller and smaller. One fears that the mathematical community will turn into a loose association of independent, incommunicative, and, ultimately, sterile technicians. Rampant specialization in itself would not undermine the unity and fertility of mathematics as long as the many narrow specialities maintained enough points of contact with one another to support a comprehensive network of essential interconnections. The unity of mathematics might then be akin to the complex unity of a living organism. But why should we suppose this will occur? "Clearly the danger is that mathematics itself will suffer the fate of splitting into different separate sciences, into many independent disciplines tenuously connected."[8]

Mathematics endures as a single science because the disintegrative trends have been counterbalanced by powerful integrative forces — forces which continually produce new and essential links between the specialized cells of the mathematical organism.

> A ... characteristic feature of modern mathematics is the formation of general concepts on a new and higher level of abstraction. It is precisely this feature that guarantees preservation of the unity of mathe-

matics, in spite of its immense growth in widely differing branches. Even in parts of mathematics that are extremely far from one another similarities of structure are brought to light by the general concepts and theories of the present day. They guarantee that contemporary mathematical methods will have great generality and breadth of application; in particular they produce a profound interpenetration of the fundamental branches of mathematics: geometry, algebra, and analysis.[9]

More specifically: *set theory* provides mathematicians from widely scattered specialities with an overarching theoretical framework which helps to reveal shared insights and concerns. For example, as Hermann Weyl points out, set theoretic considerations can establish that the introduction of points at infinity in Euclidean geometry and the development of abstract modular arithmetics in number theory are instances of a single mathematical process.[10]

This example will reward closer attention. If m, n, and p are integers and $p > 0$, then m and n are congruent modulo p (in symbols, "$m \equiv n \pmod{p}$") if and only if m and n differ by a multiple of p. So, for example, $19 \equiv 4 \pmod 3$ since $19 - 4 = 15$ and $15 = 3 \cdot 5$. We can use congruence relations to characterize arithmetical theories (known as "modular arithmetics") involving only finitely many numbers (which Weyl calls "congruence integers"). Letting, say, 13 be our modulus, we postulate that each integer n is associated with a particular congruence integer known as "$f(n)$." And we stipulate that $f(n) = f(m)$ if and only if $n \equiv m \pmod{13}$. Note that we are using a relationship between distinct members of one domain as a criterion for the identity of members of another domain. In this case, we are using the congruence of distinct integers as a criterion for the identity of congruence integers. This general technique, which is known as "definition by abstraction," will be discussed further in chapters 2 and 4. We now obtain a version of arithmetic by letting $f(n) + f(m) = f(n+m)$ and $f(n) \cdot f(m) = f(n \cdot m)$.

Turning to geometry, it is useful for certain purposes to introduce "points at infinity" at which parallel lines are said to intersect. Accordingly, we postulate that each line j is associated with a point at infinity known as "$g(j)$." And we stipulate that $g(j) = g(k)$ if and only if j and k are parallel. Here we are using the parallelism of lines as a criterion for the identity of the corresponding points at infinity. Since the point $g(j)$ is said to lie on a line k if and only if j and k are parallel, it follows that j and all lines parallel to j will be said to share the point $g(j)$. It is in this sense that parallel lines are understood to meet at an "infinitely distant point."

The insight that there are interesting arithmetics involving only finitely many numbers seems far removed from the insight that parallel lines can consistently be taken to meet. And, indeed, the first insight is based in part on an appreciation of features peculiar to number theory, just as the second is based in part on an appreciation of features peculiar to geometry. Still, the introduction of congruence integers and of points at infinity can be justified set theoretically in fundamentally similar ways – and this gives us good reason to say that these two maneuvers from widely separated branches of mathematics are themselves fundamentally similar. Let us first of all take $f(n)$, the congruence integer (mod 13) associated with n, to be the set of all integers congruent (mod 13) to n. And let us take $g(j)$, the point at infinity associated with j, to be the set of all lines parallel to j. In symbols:

$$f(n) = \{x: x \equiv n \pmod{13}\}$$
$$g(j) = \{x: x // j\}.$$

Then we can prove that $f(n)$ and $g(j)$ have the crucial properties cited above. In particular:

$$\{x: x \equiv n \pmod{13}\} = \{x: x \equiv m \pmod{13}\}$$
$$\text{if and only if } n \equiv m \pmod{13}$$
$$\{x: x//j\} = \{x: x//k\} \text{ if and only if } j//k.$$

That is, on the basis of a single group of set theoretic principles, we can establish both that the congruence of integers is a reliable criterion for the identity of the associated congruence integers and that the parallelism of lines is a reliable criterion for the identity of the associated points at infinity. We might say then that these principles express the shared content of certain construction principles from number theory and geometry (and, indeed, from many other far-flung branches of mathematics). By grasping such set theoretic principles we come to appreciate the fundamental unity of apparently unconnected specialties.

2. Mathematics' One Foundation

I have been trying to show that set theory deserves respect for its work as a mathematical adhesive. But this in itself does not sufficiently explain the loftiness of set theory's current position. For example, abstract algebra is also a powerful tool for opening channels of communication between specialists. Why then is algebra not widely hailed as "*the* foundation of mathematics"? The answer, briefly, is that set theory is both comprehensive and, in at least some of its manifestations, supremely uncontroversial. Set theorists generate widely accepted theories within which the characteristic patterns of definition and inference of all but the most heterodox mathematical subfields can be reproduced. It's one thing to uncover shared patterns of thought in a number of widely separated specialties; it's quite another to give a unified account of the abstract theoretical presuppositions of mainstream mathematics *as a whole*. Set theory does both. Thus, even while contributing to the piecemeal unification of mathematics as it is actually practiced, set theory supplies us with compelling formal reasons for regarding mathematics as a single science. After all, researchers who all can be regarded as working out the consequences of a single theory (in this case, a powerful set theory) will quite naturally be identified as practitioners of one cohesive science.

With the rise of category theory (a branch of abstract algebra), set theory's status as *the* foundation of mathematics has been seriously challenged. Category theory too opens up lines of communication between mathematicians who might otherwise ignore one another's work:

> ... the prevailing value of this theory lies in the fact that many different mathematical fields may be interpreted as categories and that the techniques and theorems of this theory may be applied to these fields. It provides the means of comprehension of larger parts of mathematics. It often occurs that certain proofs, for example, in algebra or in topology, use "similar" methods. With this new language it is possible to express these "similarities" in exact terms. Parallel to this fact there is a unification. Thus it will be easier for the mathematician who has command of this language to acquaint himself with the fundamentals of a new mathematical field if the fundamentals are given in a categorical language.[11]

So category theory, too, is a mathematical adhesive. Even more importantly, category theory can match set theory's claim to comprehensiveness: the work of mainstream mathematicians can be regarded as an exploration of the consequences of a powerful category theory. So category theorists also supply us with compelling formal reasons for viewing mathematics as a cohesive enterprise.

Since the patterns of inference characteristic of category theories can be reproduced within certain set theories, it is tempting to say that category theory is just a branch of set theory and, therefore, is not a genuine competitor. But this sort of argument cuts both ways – for the patterns of inference characteristic of set theories can be reproduced within certain category theories. So we might just as well say that set theory is a branch of category theory.

Why then is it set theory that rules the roost? The reasons are many, but I list just a few: (1) Set theory came first, got firmly entrenched in the system of mathematical education (presumably because of the virtues cited above), and has thoroughly infected the intellectual life of several generations of mathematicians right up to the present. A shift away from set theory would disrupt both the practice and pedagogy of mathematics and, hence, would have to be very powerfully motivated. Category theory flickered into being only in 1945 and has not managed similarly to pervade mathematics. (2) Since the early years of this century, the application of set theory by specialists has not, in itself, produced any mathematically significant anomalies (although set theory has contributed to some results which bruise the intuitions of laymen and philosophers). There have, in other words, been no disasters of a sort which would prompt a move away from set theory. (3) There are very embarrassing gaps in our knowledge of the set theoretic universe, but there is no prospect that category theory will help to fill these gaps. Thus, the category theoretic outlook has not demonstrated its superiority in a way that would be peculiarly compelling to set theorists. (4) Category theorists are drawn to powerful theories which are in more danger of inconsistency than the most popular set theories.[12] So, in the absence of a compelling motivation, the abandonment of those set theories seems at least mildly imprudent.

These considerations do not imply that set theory's hegemony will last forever. No single theory or theoretical tradition is likely to be immune from both anomaly and infertility. Nonetheless, set theory currently plays a singularly important role in mathematics and, so, ought to be examined thoroughly – no matter what its future may hold.

3. Science of the Infinite

Our discussion thus far might lead one to believe that set theorists are busybodies who have a hand in everyone else's affairs, but have no special interests of their own. The truth is that set theory is itself a mathematical specialty (in addition to being a glue for the other specialties). The special concern of set theorists is the development of an abstract account of the *infinite*.

Let's try to see, at an elementary level, what this involves. Imagine a classroom in which every student occupies exactly one desk and every desk is occupied by exactly one student. What can we say about the number of students and the number of desks in that classroom? Clearly, these numbers are identical. If the room contains exactly 13 desks, it must contain exactly 13 students. If the room contains exactly 105 students, it must contain exactly 105 desks. More technically, we say that the desks and the students are *equinumerous* because there is a *pairing* between them – a pairing defined by the rule: desk x is paired with student y if and only if x is occupied by y. It seems not to be merely a peculiarity of this example that the existence of a pairing implies equinumerosity and that equinumerosity implies the existence of a pairing. Rather it seems natural to state as a universal principle that objects are equinumerous if and only if they are "pair-able." On this modest foundation Georg Cantor (1845–1918), who was in many ways the "father of set theory," erected a majestic science of the infinite.

If we courageously pursue the consequences of our principle of equinumerosity, we reach some rather surprising results. For example, one might think that the natural numbers $0,1,2,3,\ldots$ are twice as numerous as the even natural numbers $0,2,4,6,\ldots$ – for the latter are the result of deleting half the former. But consider the relation R defined by:

x stands in R to y if and only if $y=2x$.

Every natural number stands in R to exactly one even natural number. And to every even natural number there stands in R exactly one natural number. So R pairs off all of the natural numbers with just the even natural numbers; and our principle of equinumerosity allows us to infer that the natural numbers and the even natural numbers are equinumerous. Using similar reasoning, we could establish the equinumerosity of the natural numbers and the multiples of 10 $(10,20,30,\ldots)$ or of 100 $(100,200,300,\ldots)$ or of 10^{100} $(10^{100}, 2\times10^{100}, 3\times10^{100},\ldots)$ or of any other number one might care to name. Perhaps more surprisingly, we could show that the points on any two lines are equinumerous – even if one of the lines is finite in length and the other extends infinitely in both directions!

These examples might lead one to believe that infinity guarantees equinumerosity. One might suppose that if there are infinitely many X's and infinitely many Y's, then the X's and the Y's are equinumerous. It is actually fairly easy to show that this is not so. (Easy *for us*. Cantor's original proof that there are distinct sizes of infinity was, at the time, a stroke of genius.)

First note that there are exactly two sequences of 0's and 1's which are one digit long: 0, 1. There are exactly four sequences of 0's and 1's which are two digits long: 00, 01, 10, 11. There are exactly eight sequences which are three digits long: 000, 001, 010, 011, 100, 101, 110, 111. And, for any non-zero natural number n, there are exactly 2^n sequences which are n digits long. But how many sequences of 0's and 1's are there which are infinitely long? More particularly, can the infinitely long zero/one sequences be paired with the non-zero natural numbers $(1,2,3,\ldots)$?

Suppose that R' is a relation such that every non-zero natural number stands in R' to exactly one infinitely long zero/one sequence. The sequence to which a natural number n stands in R' will be identified as "the nth sequence." (So, for example, the sequence to which 13 stands in R' will be identified as the 13th sequence.) Using this terminology, we now define an infinitely long zero/one sequence S as follows. If the nth digit of the nth sequence is 0, then the nth digit of S is 1. And if the nth digit of the nth sequence is 1, then the nth digit of S is 0. Suppose, for example, that the first five digits of the first five sequences look like this:

$$10100\ldots$$
$$00110\ldots$$
$$00011\ldots$$
$$11111\ldots$$
$$01001\ldots$$

Then the first digit of S is 0 since the first digit of the first sequence is 1. The second digit of S is 1 since the second digit of the second sequence is 0. And so on — the first five digits of S being 01100.

It is now easy to see that no non-zero natural number stands in R' to S. Given any non-zero natural number n, the nth digit of S differs from the nth digit of the nth sequence. For example, if we suppose (as above) that the fifth digit of the fifth sequence is 1, then the fifth digit of S is 0. So, for no non-zero natural number n, can S itself be the nth sequence. But to be the nth sequence is simply to be the sequence to which n stands in R'. We conclude, as promised, that no non-zero natural number stands in R' to S.

What have we established? We assumed merely that each non-zero natural number stands in R' to exactly one infinitely long zero/one sequence. On the basis of this piece of information alone we were able to establish the falsity of the proposition that to every infinitely long zero/one sequence there stands in R' exactly one non-zero natural number (since to S there stands in R' no non-zero natural number). But this means that every relation which has one of the essential properties of a pairing between the non-zero natural numbers and the infinitely long zero/one sequences will, for that very reason, lack the other essential property. So there are no such pairings and, hence, the non-zero natural numbers and the infinitely long zero/one sequences are not equinumerous. (By the way, this piece of reasoning is known as a "diagonal argument" because of the diagonal pattern formed by the digits printed in bold-face above.)

We see, then, that when dealing with an infinitude of objects there is more than one possible answer to the question "How many?" The response "Infinitely many!" might once have seemed adequate. But since the number of one infinitude of objects can differ from the number of some other infinitude of objects (i.e., the former objects can fail to be equinumerous to the latter), we would like to know what number *in particular* such objects might have.

By introducing a scale of infinite cardinal numbers, Cantor showed that determinate answers to "How many?" questions involving infinitely many objects can be given. Using Cantor's terminology, we say that the natural numbers have the cardinal number $\aleph_0$ — or, equivalently, the natural numbers are $\aleph_0$ in number. ('$\aleph$' is the Hebrew letter aleph.) By showing that mathematically interesting operations (such as addition and multiplication) can be applied to the infinite cardinals, Cantor also became the founder of an infinite cardinal arithmetic. Of particular interest in this connection is the operation of exponentiation. Not only can we define the expression '$2^{\aleph_0}$', we can even show that the infinitely long zero/one sequences are precisely $2^{\aleph_0}$ in number. Furthermore, any infinite cardinal $\aleph_\alpha$ is demonstrably smaller than 2 raised to the power of $\aleph_\alpha$ — i.e., $\aleph_\alpha < 2^{\aleph_\alpha}$. If we let $\beth_0 = \aleph_0$ and $\beth_{\alpha+1} = 2^{\beth_\alpha}$ ('$\beth$' being the Hebrew letter beth), then we are guaranteed an infinitely increasing sequence of infinite cardinals: $\beth_0 < \beth_1 < \beth_2 < \ldots$ This means that there are at least as many infinite cardinal numbers as there are finite ones. (And, in fact, there are *many more* infinite than finite cardinals.)

Cardinals are not the only numbers we recognize. When we say that there are nine players on a baseball team we are giving a possible answer to a "How many?" question. But when we say that Dick Green bats ninth we are doing no such thing. We are not assigning Dick Green (or anything else) a magnitude, but rather indicating his position in an ordering (in this case, a batting order). Numbers which are used to indicate such positions are known as "ordinals." We are all familiar with the finite ordinals (1st, 2nd, 3rd, 4th, 5th,...). It took the genius of Cantor to show that the notion of ordinal number can be extended into the *infinite*.

Let $\ll$ be a non-standard ordering of the natural numbers defined as follows. If n and m are either both even or both odd, let $n \ll m$ if and only if $n < m$ (where $<$ is the standard "less than" relation). If n is odd and m is even, let $n \ll m$. Thus $\ll$ orders all the odd numbers in the usual way (with respect to each other) and similarly all the even numbers, but it places all the even numbers *after* all the odd ones. 1 is the 1st natural number in this ordering, 3 is the 2nd, 5 is the 3rd, and so on. But what is 0? In Cantor's terminology, 0 has the ordinal number ω in this ordering. ('ω' is the Greek letter omega.) The ordinal number of 2 in this ordering is $\omega+1$, the ordinal number of 4 is $\omega+2$, and, more generally, the ordinal number of $2n$ is $\omega+n$. (A word of warning: '+' here represents a special ordinal operation which differs from cardinal addition in ways we shall not explore.)

ω is the first infinite ordinal number. Note that the ordinal numbers which precede ω (i.e., the finite ordinals 1st, 2nd, 3rd, ...) and the natural numbers which precede 0 in our non-standard ordering (i.e., the odd natural numbers 1, 3, 5, ...) form sequences which, from a structural point of view, match up perfectly:

$$1, \quad 3, \quad 5, \quad 7, \quad 9, \ldots$$
$$\downarrow \quad \downarrow \quad \downarrow \quad \downarrow \quad \downarrow$$
$$\text{1st, 2nd, 3rd, 4th, 5th,} \ldots$$

Similarly, the predecessors of $\omega+1$ form essentially the same structure as the $\ll$-predecessors of 2:

$$1, \quad 3, \quad 5, \quad 7, \quad 9, \ldots ; \quad 0$$
$$\downarrow \quad \downarrow \quad \downarrow \quad \downarrow \quad \downarrow \quad \quad \downarrow$$
$$\text{1st, 2nd, 3rd, 4th, 5th,} \ldots ; \quad \omega$$

More generally, the predecessors of $\omega+n$ form essentially the same structure as the $\ll$-predecessors of $2n$:

$$1, \quad 3, \quad 5, \ldots ; \quad 0, \quad 2, \quad 4, \ldots, \quad 2(n-1)$$
$$\downarrow \quad \downarrow \quad \downarrow \quad \quad \downarrow \quad \downarrow \quad \downarrow \quad \quad \downarrow$$
$$\text{1st, 2nd, 3rd,} \ldots ; \quad \omega, \omega+1, \omega+2, \ldots, \omega+(n-1)$$

And the predecessors of the finite ordinal "nth" form essentially the same structure as the $\ll$-predecessors of $2n-1$:

$$1, \quad 3, \quad 5, \ldots, \quad 2(n-1)-1$$
$$\downarrow \quad \downarrow \quad \downarrow \quad \quad \downarrow$$
$$\text{1st, 2nd, 3rd,} \ldots, \quad (n-1)\text{th}$$

These observations suggest that ordinals can be used to characterize whole structures. (We needn't use them *just* to identify individual positions within structures.) The trick is to assign an ordinal α to a structure X if and only if the predecessors of α form a structure which is *isomorphic* to X. An isomorphism is a pairing which preserves relative position within structures. A pairing which matched 0 with ω+1 and 2 with ω would not be an isomorphism with respect to ≪ and the usual ordering of the ordinals. For 0 occurs earlier than 2 in the ≪ ordering, while ω+1 occurs later than ω in the usual ordering of the ordinals. On the other hand, the relation R defined above (which pairs off each natural number n with $2n$) *is* an isomorphism between the natural numbers and the even natural numbers with respect to the usual ordering of the naturals. For $n < m$ if and only if $2n < 2m$. So R provides us with the following structural match:

$$0, 1, 2, 3, 4, \ldots$$
$$\downarrow \downarrow \downarrow \downarrow \downarrow$$
$$0, 2, 4, 6, 8, \ldots$$

The predecessors of ω and the ≪-predecessors of 0 form structures which are isomorphic in this precise sense. (Just match up each finite ordinal "nth" with the odd natural number $2n-1$.) So, employing our trick for tagging structures with ordinals, we assign the ordinal ω to the infinite sequence 1,3,5,... To take another example, since the predecessors of ω+1 and the ≪-predecessors of 2 form isomorphic structures, we assign the ordinal ω+1 to the ordering 1,3,5,...; 0. And, more generally, since the predecessors of ω+n and the ≪-predecessors of $2n$ form isomorphic structures, we assign the ordinal ω+n to the ordering 1,3,5,...; 0,2,4,..., $2(n-1)$.

Ordinals can be assigned to every structure which is sequence-like in the sense of being *well-ordered* (a notion to be defined in chapter 8). Two well-ordered structures X and Y which are isomorphic to one another will always be assigned the same ordinal. For example, when ordered by the standard "less than" relation, the natural numbers and the even natural numbers form isomorphic well-orderings. Each of these well-orderings is assigned the ordinal ω, because each is isomorphic to the structure formed by the predecessors of ω:

$$0, \quad 1, \quad 2, \quad 3, \quad 4, \ldots$$
$$\downarrow \quad \downarrow \quad \downarrow \quad \downarrow \quad \downarrow$$
$$0, \quad 2, \quad 4, \quad 6, \quad 8, \ldots$$
$$\downarrow \quad \downarrow \quad \downarrow \quad \downarrow \quad \downarrow$$
$$\text{1st, 2nd, 3rd, 4th, 5th,} \ldots$$

Since the isomorphism of well-orderings is a reliable criterion for the identity of the associated ordinals, we might say that ordinals are abstract objects which are linked to isomorphic well-orderings in just the same way that points at infinity are linked to parallel lines and congruence integers are linked to congruent integers.

4. But What Are Sets?

We have had much to say about set theory as a mathematical discipline, about the general characteristics of individual set theories, and about the concerns of set theorists. We have seen that set theory functions as a mathematical adhesive, that set theories uniquely combine comprehensiveness and lack of controversy, and that set theorists are specialists whose particular interest is the infinite. Yet we have had virtually nothing to say about *sets* themselves. Isn't set theory the theory of sets? And so, this being a treatise on set theory, are we not entitled to be told what sets are?

Some readers may already have decided that I am taking a basic acquaintance with mathematical sets for granted — that I am assuming that most everyone already knows what mathematical sets are. That, at least, would explain why I have not defined the notion "set." But, in fact, this hypothesis could not be further from the truth.

I anticipate that virtually no one reading this book will already be in a position to say what mathematical sets are. Indeed, the following two chapters will reveal the folly of supposing that mathematical sets are objects with which people are typically acquainted. Having undermined this mistaken view, the more positive task of determining what mathematical sets might be or, more neutrally, explaining what set theory might actually be about will occupy us for the remainder of this book.

II
OFFSPRING OF ANALYSIS

> ... the knowledge of one generation of mathematicians is obtained by extending the knowledge of the previous generation. To understand the epistemological order of mathematics one must understand the historical order. — Philip Kitcher[13]

> Under the present dominance of formalism, one is tempted to paraphrase Kant: the history of mathematics, lacking the guidance of philosophy, has become *blind*, while the philosophy of mathematics, turning its back on the most intriguing phenomena in the history of mathematics, has become *empty*.
> — Imre Lakatos[14]

1. A Role for History

Philosophers of science have learned from hard experience that contributing to the *mythology* of science is generally the price of ignoring the *history* of science. Abstract philosophical accounts of science which are formulated without the benefit of any inquiry into how concrete communities of scientists actually behave will, most likely, get things very wrong. Philosophers who are thus disengaged from their subject matter have even been known to formulate methodological rules which, if followed, would bring science to a screeching halt.

It has been a great blessing that the interests of some enlightened philosophers of science, such as Imre Lakatos and Philip Kitcher, have extended into the mathematical domain. For these thinkers have ably carried the gospel of historical research into the philosophy of mathematics. Mathematics, after all, is the product of communities of human inquirers. A mathematical proposition can only have a meaning which is conferred on it by historically situated users of mathematical languages. A mathematical theorem can become known only through the application of techniques which can be wielded by flesh and blood mathematicians. Hence, expositions of what mathematicians actually do and have done would surely be of value to philosophers concerned with the meanings which mathematical formulas can bear and with the truths which mathematicians can know.

We see, then, that there are reasons of a quite general sort for philosophers of mathematics to educate themselves about the history of mathematics. But one would still like to know why, more particularly, a philosopher of *set theory* should become historically literate (for it is the aim of this chapter to contribute to precisely such literacy). The first response to this query which comes to mind is, again, quite general in character: a survey of set theory's historical genesis would be valuable even if it did no more than remind us that set theory *has* a history; that in itself would aid the nascent historicism in the philosophy of mathematics. I hasten to note, however, that I also have a more narrowly focused philosophical goal. I would like to undermine two myths which stand in the way of an adequate philosophical grasp of set theory.

Myth #1: The development of mathematical set theory has been significantly influenced by notions borrowed directly from everyday thought.

Myth #2: One can be expected to have an essentially sound notion of what a mathematical set is prior to learning anything about mathematical set theory.

My primary target is Myth #2. The apparent familiarity of mathematical sets is a dangerous illusion which must be dispelled by philosophical therapy before any fundamental progress in the philosophy of set theory can be made. (We must first know that we do not know!) Myth #1 is to be abhorred because, in addition to being untrue, it helps to prop up Myth #2. I don't actually know whether anyone self-consciously believes Myth #1. But I should think anyone who self-consciously *disbelieves* it would be skeptical of Myth #2. (If mathematical set theory was not significantly molded by commonsense notions, then why should the mere possession of such notions be expected to familiarize anyone with mathematical sets?) So we can say that the lack of *disbelief* in Myth #1 helps support Myth #2. In the remainder of this chapter, I hope to cultivate just such disbelief.

Note that the myths cited above involve *mathematical* sets. I readily admit that there are mundane set concepts with which every speaker of English is conversant. Indeed, we shall consider just such notions in the following section. So I am not claiming that sets *of all types* are alien to the mathematically uninitiated. The point is merely that an acquaintance with mundane sets does not guarantee even a minimal grasp of mathematical sets.

2. Greek Roots

We shall see that set theory as we now know it grew out of the nineteenth-century theory of functions and real numbers (a theory known, more briefly, as "analysis"). Since the Greeks failed to admit the existence of most real numbers, the ancient groundwork for modern analysis is to be found primarily in the Greek theory of space rather than in their theory of number. And, indeed, it was as geometers that Eudoxus and Archimedes approached the calculus of Leibniz and Newton. Nonetheless, Greek number theory merits our attention for we find there an unusually close alliance between formal science and commonsense notions of set. To ignore Greek number theory would be unfairly to ignore one of Myth #1's most likely sources of support.

The ancient Greek study of number was divided into two subdisciplines: arithmetic (ἀριθμητική) and logistic (λογιστική). Very roughly, arithmetic concerned itself with the classification of numbers into various species (εἴδη), whereas logistic was concerned with techniques of calculation. So, for example, the young Theaetetus displays a knack for arithmetic when he proposes that "numbers be divided into two classes: the square and the oblong" (*Theaetetus* 147E). On the other hand, someone who is adept at calculation (someone who is among οἱ φύσει λογιστικοί – *Republic* 526B) is especially likely to contribute to logistic.

Arithmetic and logistic were thought to differ not only in their objectives, but also in their objects. An anonymous *scholium* to Plato's *Charmides* 165E tells us that logistic deals not with numbers (as arithmetic does) but with numerables (λογιστική ἐστι θεωρία τῶν ἀριθμητῶν, οὐχὶ δὲ τῶν ἀριθμῶν μεταχειριστική).[15] And Theon of Smyrna, a Platonist of the second century A.D., tells us what this distinction amounts to: a number (ἀριθμός) is an "intelligible quantity, such as 5 itself or 10 itself" (τὸ ἐν νοητοῖς ποσόν, οἷον αὐτὰ ε′ καὶ αὐτὰ ι′); on the other hand, a numerable (ἀριθμητόν) is a "perceptible multiplicity, such as 5 horses or 5 cows or 5 people" (τὸ ἐν αἰσθητοῖς ποσόν, ὡς ἵπποι ε′, βόες ε′, ἄνθρωποι ε′).[16]

Numerables are simply pluralities of concrete objects. We can refer to them using either plural expressions such as 'these horses' or singular expressions such as 'this team of horses'. Thus, they are objects of the sort of set or collectivity talk which occurs in everyday nontechnical language – they are commonsense sets. (That is, we can make good sense of *ancient Greek* thinking about numerables by employing a notion of set which is commonsensical *for us*. I am not claiming that what is commonsensical for us was commonsensical for the Greeks.) Numerables are collections, groups, suites, classes, ensembles, species, flocks, bands, packs, hordes, throngs, coteries, assemblages, multitudes, gatherings, complements ("a full complement of teeth"), companies ("the glorious company of the apostles"), parties ("Burton's party approached the summit"), outfits ("Corporal Glinka rejoined his outfit"). They are numbers *in the colloquial English sense* (as in, "A number of sheep played cards"). Of course, a number in Theon's sense is something rather different.

A number, like a numerable, is a plurality ($\pi\lambda\tilde{\eta}\theta o\varsigma$). For Theon, a "one" ($\tilde{\varepsilon}\nu$) is the fundamental constituent of a numerable, whereas a unit ($\mu o\nu\acute{\alpha}\varsigma$) is basic to a number. Ones are just concrete individuals. The ones of a particular numerable are just the objects of which that numerable consists. For example, the ones of the twelve apostles are each of the twelve apostles. On the other hand, since a number is an abstract representation of a class of equinumerous numerables, the units which make up numbers are not concrete individuals, but abstract representations of such individuals. (Aristotle says quite explicitly that mathematical objects are products of an abstraction process. Cf. *Metaphysics* 1061a.) We might say that the number 5 is a plurality which has only those characteristics shared by all five-membered numerables. A unit, then, would be a concrete individual which has been drained of all content apart from its discreteness from other individuals.

The ones of a numerable are indivisible with respect to that numerable. For example, if we intend to count and jointly refer to the twelve apostles, then we deal with whole apostles – we don't pick out and count parts of them. But whereas a one is a one and is indivisible only relative to a particular act of designation and counting, a unit, in its abstractness, has no such relativity and, thus, can be declared absolutely indivisible. So, to say that the ones of a numerable are indivisible is just to say that when we designate and count objects of a particular type t, we designate and count whole t's. On the other hand, to say that the units of a number are indivisible in some non-relative sense is just to say that when we refer to units, we abstract from every particular type t and deal with objects which are (or appear to be) free of all characteristics other than their mutual discreteness. Thus, the objects of logistic are simply concrete individuals taken plurally (that is, they are commonsense sets), while the objects of arithmetic are, if we follow Aristotle, pluralities of those very same concrete individuals which, however, have been "denatured" through a severe process of abstraction.

I have already mentioned that most of the ancient foundations for modern analysis (and, thus, for contemporary set theory) were laid by Greek geometers rather than by Greek number theorists. So we should not expect to find any very strong historical link between Greek logistic notions and modern set theoretic ones – in spite of logistic's being what we today would regard as a theory of commonsense sets. Nonetheless, Jacob Klein has identified a very weak link between logistic and set theory.[17] Both directly, through the fifteenth-century reemergence of Diophantus' logistical work, and indirectly, through the influence of Diophantus on Arabic mathematics, Greek logistic helped to shape modern mathematics. However, in the course of the modern assimilation of logistical results, the Greek notions of number and numerable were replaced by radically different conceptions. So the close alliance between Greek number theory and commonsense set talk was lost and played no significant role in the subsequent development of mathematical set theory. Although, as Klein has shown, Greek logistic can be regarded as a precursor to contemporary set theory because of its general influence on modern mathematics, the characteristic of logistic which has most interested us (namely its preoccupation with what we regard as commonsense sets) contributed in only a very minor way to the nineteenth-century interest in the notion of set; and it did not significantly help to shape modern mathematical treatments of this notion. So we must look elsewhere for illumination about the development of modern set theory.

3. The Latitude of Forms

Since, as I have already mentioned, contemporary set theory arose from nineteenth-century analysis, we might do well to explore the origins of that mathematical discipline. Accordingly, we turn from ancient Greece to late medieval England. For the modern theory of real-valued functions owes much to the seminal work of the so-called "Oxford Calculators" of the fourteenth century – most notably, Thomas Bradwardine, William Heytesbury, Richard Swineshead, and John Dumbleton. If the medievalist Edith Dudley Sylla is to be believed, the Calculators were, regrettably, motivated less by a lust for knowledge than by the all too familiar need to keep undergraduate logic students well supplied with suitable exercises and exam questions—the unfortunate result being that the most profound insights take on, at the hands of the Calculators, the form of curious and rather awkward puzzlers.[18] Be that as it may, their style and motives are really of no concern to us. What matters is that they identified a range of problems which prompted some important reflections on what we today recognize as the notion of functional dependence.

The Calculators were particularly interested in the "latitude of forms." That is, they particularly concerned themselves with variations in physical attributes (such as temperature and velocity) which are subject to what we would regard as continuous increase and decrease in intensity. The medium within which this increase or decrease takes place was known as the "extension" of the variation (and would, most naturally, be either a time interval or a spatial region). So the variation of a form could be (and, more importantly, was) thought of as a correlation between the intensities attained in the course of the variation and the components of an extension—each such component being correlated with exactly one intensity. This means that whether or not the Calculators possessed any abstract notion of mathematical function, they were most certainly involved in the mathematical representation and treatment of particular functional relations.

The most impressive achievement of the Oxford Calculators was their derivation of the kinematic law of uniformly accelerated motion (a law "discovered" by Galileo about two and a half centuries later). Suppose some attribute increases or decreases in intensity at a uniformly accelerating rate over some linear region of time or space. Then, as the Calculators showed, the average intensity of the attribute within that region is the mean of its intensities at the beginning and end of the region. For example, letting the attribute under consideration be velocity, the distance traveled by a uniformly accelerating object X over a certain time interval is identical to the distance that would be traversed over that same interval by an object whose velocity is the mean of the initial and terminal velocities of X. In the case of a uniformly accelerating object which begins the time interval at rest, we get the equation

$$s = \tfrac{1}{2} v \cdot t$$

where s is the distance traversed, v the terminal velocity, and t the length of the time interval. This is equivalent to Galileo's

$$s = \tfrac{1}{2} a \cdot t^2$$

where a is the rate of acceleration. For $v = a \cdot t$.

These results led Nicole Oresme, an important French mathematician of the fourteenth century, to a startling (though only partial) anticipation of analytic geometry. Oresme realized that if an extension were represented by a horizontal line, then vertical lines erected on points of the horizontal one could represent the intensities attained by an attribute at the corresponding points of the extension. Since Oresme let the length of the vertical lines represent the degrees of intensity attained by the attribute, the curves composed of the endpoints of these lines depicted the functional relation between intensity and extension in the way familiar to us today. For example, the line MP in the diagram below would represent an attribute whose intensity remains constant throughout the interval OE. On the other hand, the line OI would represent an attribute whose intensity increases from zero at a uniformly accelerating rate. If the attribute at issue here were velocity, then the area of the rectangle $OMPE$ (being the constant velocity OM times the length of the time interval OE) would represent the distance traveled within OE by an object with velocity OM. Similarly, the area of the shaded triangle OIE can be taken to represent the distance traveled within OE by an ob-

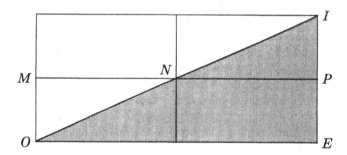

ject starting from rest and uniformly accelerating to velocity EI. But if, as in the diagram, OM is exactly half of EI, then the area of OEI is identical to the area of $OMPE$ (since the triangles OMN and NIP are congruent). So the distance traversed within OE by the uniformly accelerating object would be one half its terminal velocity times the length of the time interval – as the Oxford Calculators had previously (but much less elegantly and suggestively) established.

To sum up then: The Oxford Calculators of the fourteenth century interested themselves in physical processes whose mathematical treatment invited reflection on the notion of functional dependence. This led one of their contemporaries, Nicole Oresme, to develop the rudiments of the technique of graphically representing functions (a technique which played an extremely important role in subsequent thought about the concept "function"). Oresme's work was to be complemented, two centuries later, by François Viète's contributions to the modern symbolic/algebraic representation of functions.

4. Eidetic Logistic

Under the influence of Greek logistic and Arabic algebra (the latter being itself influenced by the former), Viète developed a *logistice speciosa* (a calculus of forms or "eidetic logistic"). His key insight was that an art of calculation could be developed whose objects were neither determinate pluralities of concrete objects nor denatured, but nonetheless cardinally determinate, pluralities of units. Viète realized that logistic could adopt an even more abstract perspective, concerning itself with the universal properties which numbers have regardless of their particular position in the number sequence.

As Jacob Klein has shown, this new perspective precipitated a radical transformation of the Greek notions of ἀριθμητόν and ἀριθμός.[19] Viète's work helped to undermine the Greek view that numbers are either pluralities of given objects or abstract representatives of such pluralities. And it helped to foster the modern presumption that our access to numbers somehow depends on our mastery of techniques of symbolic manipulation. Thanks in part to Viète, we no longer regard the plural ostension of concrete individuals as the fundamental mode of acquaintance with the objects of logistic (or, as we would say, algebra). Instead, logistic itself has become the primary vehicle for acquaintance with its own objects.

That we would today hesitate to ascribe knowledge of the natural numbers to anyone inept at arithmetical calculation and unable to solve number theoretic problems is a sign of how little hold the Greek concept of number has on us. From the Greek point of view, to be conscious of several objects at once is to be acquainted with a numerable. But to be acquainted with numerables is to be only one act of abstraction away from a minimal knowledge of numbers. So, just as someone innocent of ornithology can nonetheless be well acquainted with birds, someone unschooled in arithmetic and logistic need not be blind to pluralities of "ones" or to their abstract counterparts, pluralities of units. By contrast, while we today might not be terribly clear about what numbers are, we do seem fairly confident about closely linking knowledge of them with mastery of arithmetical techniques.

The move away from ἀριθμητόν and ἀριθμός opened the way for the algebraic treatment of numbers other than the naturals. It was not until the fourth century A.D. that a Greek ventured to regard fractions as numbers in their own right rather than as ratios. And even this brave soul (Diophantus) would have no truck with equations whose roots are negative or irrational. Yet once the notion that numbers are pluralities of units ceases to hold sway (as it did, I should note, among the Arabs well before Viète), the "weirdness" or "impossibility" of such equations fades. For negatives and irrationals are "number-like" in precisely the way which has come to be of the greatest importance—that is, they are subject to algebraic manipulation and, more positively, play a useful role in the systematization of our mathematical calculi.

To say that Viète contributed to the development of our modern concept of number is not to say that this was one of his aims or that he had an adequate grasp of the transformation being undergone by classical conceptions. Indeed, Viète probably would have vigorously denied having any revisionary intent. Similarly, Viète might never have reflected upon the abstract notion of function. But, nonetheless, his contribution to the theory of functions ranks with that of Oresme. Whereas Oresme helped to pioneer the graphical depiction of functions, Viète's work facilitated their symbolic representation.

It is basically correct to credit Viète with introducing the modern western mathematical world to the algebraic manipulation of variables. But we must be careful to understand exactly what this means. Twelve centuries prior to Viète, Diophantus was solving equations involving an "unknown number" (ὁ ἀόριστος ἀριθμός) and, so, might well lay claim to the algebraic employment of variables. However, it would be more accurate to say that Diophantus' sign for an unknown played a mathematical role which is today often filled by signs which happen also to be used as variables—which, of course, is very different from saying that Diophantus employed variables. An example will help to make this clearer. Suppose we are faced with the mathematical word-problem: "Find the number whose square is seven less than the square of its successor." It is generally illuminating to write out the information contained in a word-problem in the form of an equation, letting a variable stand for the number to be found. Thus we have:

$$(x+1)^2 - x^2 = 7.$$

This quickly yields the result: $x = 3$. In our equation, the variable 'x' serves as a name for an initially unknown but, nonetheless, determinate number. The word-problem is solved when we discover what number 'x' names. (In this case, 3.) In English we might say: "There is a unique number which has such-and-such a property; for the moment, let's call that number 'x'." We might just as well say, "Let's call it 'Diophantus'" or "Let's call it 'François Viète'." The point is to make our task easier by bestowing a name on a definite number. That we should use variables for this purpose is little more than an historical accident. By contrast, in the algebraic formula

$$(x+1)^2 = x^2 + 2x + 1$$

the variable 'x' does not name any particular number. Rather it functions as a kind of pronoun linked to a suppressed universal quantifier: "Given any real number x, $(x+1)^2 = x^2 + 2x + 1$." Whereas Viète emphasized the algebraic manipulation of variables such as these (variables which do not stand for particular determinate numbers), Diophantus' primary concern was the application of logistic techniques to letters serving as names for unknowns. Of course, this doesn't mean that the Greeks were unaware of or were unable to express mathematical truths which for us correspond to universal generalizations with bound variables. The point is merely that they failed to emphasize an algebra within which variables ranging indifferently over some domain of mathematical objects are manipulated for purposes other than the discovery of particular numbers.

Although Viète's algebra is relatively close to our own, we should not assume that his conception of variables is the one familiar to us. Viète regarded his algebra as a *logistice speciosa*: a calculus of forms or species. Unlike classical logistic, this discipline was to concern itself with numbers only indirectly. According to Jacob Klein, its primary objects, the forms of arithmetical or geometrical magnitudes, were to occupy a higher level of abstractness.[20] Whereas we regard the variables of an algebra as pronominal expressions which range over whatever domain we happen to supply, there is some reason to believe that Viète's variables were meant to refer directly to special abstract magnitudes which are determinate only with respect to their degree (that is, "magnitudes" which, having no particular size, are identifiable only as "sides" or "squares" or "cubes" or "squared squares," etc.). Thus the ability of Viète's variables to express generality would not be a product of their own syntactic and semantic properties, but rather of the inherent abstractness of the objects they designate.

To be honest, it is not really clear how Viète interpreted his variables. (He seems at times to suggest that they are no more than meaningless place-holders.) Still, this much seems uncontroversial: by employing variables as signs of generality rather than as names of unknowns, Viète paved the way for the modern symbolic representation of functions. If 'x' is taken to be a name of a particular unknown number, then, for example, the expression

$$3x^2+5$$

is simply the name of another unknown but, nonetheless, definite number. But once the variable 'x' is taken as a mark of generality or as a sign for an entire species of numbers, the compound expression '$3x^2+5$' itself suggests a multiplicity of numbers. And one might then be led to consider the relation between the individual mathematical objects which somehow fall under 'x' and those which fall under '$3x^2+5$'. Thus is born the algebraic treatment of functions.

5. *Genitum* and *Functio*

Although Oresme and Viète pioneered the two currently standard techniques for representing functions (and although René Descartes and Pierre de Fermat then united the graphical and algebraic approaches in their analytic geometries), it appears that no one reflected in a fully self-conscious way on an abstract notion of function prior to Isaac Newton and Gottfried Wilhelm Leibniz. Had Newton and Leibniz not so reflected, it is hard to imagine how they could have welded their predecessors' scattered results about differentiation and integration into a single calculus. For an appreciation of differentiation and integration in the abstract would seem to require some sort of recognition that they are operations which map functions into functions and, hence, would seem to require some grasp of the notion "function" itself. This is a nice example of a close link between the development, on the one hand, of a highly general mathematical theory and, on the other, of a highly abstract mathematical concept.

Newton's conception of function was, naturally enough, not the one familiar to us today. But, nonetheless, in Newton's notion of a *genitum* or "generated quantity" one can easily recognize an ancestor of our own concept of function. A *genitum* is a quantity whose "flux" can be algebraically expressed in certain prescribed ways. Whereas we would say that an expression such as

$$x^2 - 2x$$

represents a single function, Newton would say only that each term taken separately represents a *genitum*. That is, 'x^2' and '$2x$' each represent a *genitum*, but 'x^2-2x' does not. To state the point more generally: while multiplication, division, and the extraction of roots are all admitted in the production of a *genitum*, addition and subtraction are excluded. This restriction begins to make some sense when we notice that Newton introduces the notion of *genitum* in a section of *Principia* devoted to the effects of differentiation on various algebraic operations (cf. *Principia*: Book II, Lemma II). From this point of view, addition and subtraction (being unaffected by differentiation) are of little interest and might well be excluded from the discussion. Be that as it may, the important thing is that Newton explicitly considers an abstract notion of functional dependence. That we regard his account of functionality as overly narrow and somewhat mysterious does little to diminish the importance of his conceptual breakthrough.

Newton also contributed to the transformation (begun by Viète) of the classical concept of number. John Wallis, Newton's most important immediate predecessor in English mathematics, conceived of a number as "less a collection of units than an abstract ratio of any quantity to another, a definition which also includes irrational ratios as numbers."[21] Newton went even further, bestowing the title of 'number' even upon negative ratios. Thus Newton can be credited with an important role in the evolution of both the concepts whose histories we are tracing: number and function.

We appear to owe our use of the word 'function' (Latin, '*functio*') to that master of terminology and notation, Leibniz. In this connection, we would do well to remember that a concept can be grasped long before a name is coined for it. While Leibniz' terminological innovation came rather late in his career (c. 1692), it was preceded by many years of reflection on the notion of functional dependence. In fact, Leroy Loemker detects Leibniz' concern with functionality not only throughout his mathematical work, but even in his philosophical thought (particularly in his doctrine that the current state of a monad "expresses" the current states of all other monads – the relation of "expression" being akin to functional dependence).[22]

Leibniz himself applied the term '*functio*' rather narrowly to certain relations between geometric objects. For example, he referred to a curve's tangents as functions of the curve. Although Jean Bernoulli acquired '*functio*' directly from Leibniz, he associated it less closely with geometry than had its creator. Like Newton, Bernoulli linked functions with algebraic expressions, each function being denoted by a scheme containing exactly one variable (e.g., '$18x^5-7$'). Displaying some of his mentor Leibniz' skill at forging new notation, Bernoulli used capital letters to denote functions of the corresponding lower case variables (e.g., letting $X=5x$). Approaching our own usage even more closely, he also used 'φx' to denote functions of x. We have Jean Bernoulli's pupil Leonhard Euler to thank for the familiar '$f(x)$'. In these notational innovations we can detect an appreciation of Viète's logistic variables as aids in the treatment of functions.

6. Functions as Notation

According to Carl Boyer, the differential and integral calculi were initially viewed as little more than handy tools for the solution of geometric problems.[23] Although we have credited Newton and Leibniz with a grasp of the concept "function" which was notably firmer than that of their predecessors, neither was sufficiently free of the long-standing prejudice for geometry to set up a theory of functions as an independent rival to the theory of space. This task was left for Euler.

Euler had a brilliant appreciation of functions as mathematical objects worthy of special study. And he was in many ways the founder of what we now know as analysis. But, oddly enough, he also played an important role in popularizing an unfortunate and, ultimately, unacceptable account of what functions are. Whereas Newton seems to have thought that functions are *denoted by* certain mathematical expressions, Euler states quite explicitly that functions simply *are* such expressions – that is, he proposes that functions be regarded not as relations between mathematical objects, but as notations in a mathematical language. Euler was extremely liberal with regard to the operations he allowed in the construction of his functions (admitting, for example, sums of infinite series). So his conception of function was, in fact, much broader than Newton's notion of a *genitum*. However, the general idea of functions as relations is ultimately less restrictive than Euler's idea of functions as items in a language.

We today would say that the number of real-functions-as-linguistic-expressions is tiny in comparison with the number of real-functions-as-relations. (In fact, they differ by at least two orders of infinity.)

Euler's nominalist definition of 'function' is to be found in chapter 1 of his textbook *Introductio in analysin infinitorum*, written in 1745. In the preface to his 1755 *Institutiones calculi differentialis*, he is at pains to construe functional dependence in the broadest possible way—his commitment to functions-as-expressions appearing to have entirely vanished. Nonetheless, Euler's original, narrower view retained its influence both over his contemporaries (Condorcet being a notable exception[24]) and, indeed, over Euler himself. For example, Joseph Louis Lagrange, in his 1797 *Théorie des fonctions analytiques*, states as a matter of course that functions are mathematical expressions. And Euler, in the *Institutiones* itself, avers that a graph represents a function only if it can be captured by a single analytical expression (that is, a single expression built up out of the basic linguistic resources of analysis: algebraic symbols, trigonometric and logarithmic signs, limit-taking operators, infinite sum and product signs, symbols of differentiation and integration, etc.). He thus implies that, while functions and analytical schemes may not be identical to one another, they at least stand in one-to-one correspondence.

To Euler's credit, this view did not prevent him from studying and exploiting piecewise-smooth curves which (as in the diagram below)

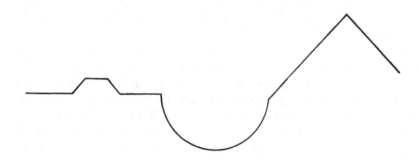

seem to correspond to a patchwork of analytical expressions rather than to a single such expression. Thus, in the celebrated controversy over the best mathematical representation of a vibrating string, Euler is sanguine about admitting such curves as initial positions of the string, whereas Jean D'Alembert hesitates, fearing that these curves "surpass the power of known analysis" (since they are not continuously differentiable).[25] A half-century later, Euler's boldness was decisively vindicated (while his distinction between genuine functions and mere patchworks of such functions was utterly exploded) by Joseph Fourier's announcement that, over a finite interval, *every* piecewise-smooth curve with only a finite number of discontinuities can be captured by a single analytical expression. (To be honest, Fourier himself would not have formulated his insight in this way. But, nonetheless, this is probably close to what he intended.)

Fourier's technique of analytical representation was partly anticipated by Jean Bernoulli's son Daniel who, in 1753, used trigonometric series to express the motion of vibrating strings. In the course of his 1807 memoir on heat diffusion, Fourier argued, in essence, that *every* periodic graph can be represented analytically by at least one trigonometric series. That is, each ordinate $g(x)$ of a periodic graph corresponds to the sum of infinitely many terms of the form '$a_n \sin nx$' or '$b_n \cos nx$'. This means that, contrary to eighteenth-century preconceptions, every graphically representable (but not necessarily either continuous or differentiable) curve whose abscissa extends from, say, 0 to 2π depicts an Eulerian, nominalistic function (restricted to that same domain).

Like Euler in his *Institutiones*, Fourier in his *Théorie analytique de la chaleur* of 1822 gives an extremely general characterization of the concept "function," insisting that the correspondence between abscissa and ordinate can be utterly anarchic. Nonetheless, as Charles Edwards points out, Fourier confined himself to well-behaved, readily visualized functions.[26] Peter Lejeune Dirichlet took Fourier's anarchistic view more seriously than did Fourier himself. A symptom of this is the birth, in 1829, of the famous everywhere-discontinuous (and utterly unpicturable) Dirichlet function f such that $f(x)=1$ if x is rational, $f(x)=0$ otherwise. Dirichlet's ultra-libertarian approach to functions opened the way for a host of such highly unintuitive and ill-behaved "curves" whose mathematical treatment demands a grasp of large and complex structures of points. Accordingly, Georg Cantor, in his own work on Fourier series, applied himself to the study of these structures.

7. Cantor the Analyst

The common claim that Cantor "invented set theory," though basically correct, is apt to be misunderstood. The rudiments of our current set theories did not spring from his head fully formed nor did they spring from his head alone. In fact, in his earliest work, in which he lays the foundations for our current set theories, Cantor betrays little interest in sets *per se*. Whereas Bertrand Russell, for example, was drawn to the very notion of "set" or "class" in his search for a reliable foundation for analysis, Cantor was drawn to huge infinite structures (which we would only now label "set theoretic") because they promised to be useful tools *within* analysis. Joseph Dauben has revealed that if Cantor was interested in sets, it was not, at least initially, as the founder of an autonomous discipline to which analysis might be reduced, but rather as a researcher eager to solve puzzles internal to analysis.[27]

The most important puzzle to which the young Cantor turned his attention involved Fourier series. In 1829, Dirichlet gave an impressive partial solution to the Fourier series convergence problem. He showed that an extensive class of functions (in his broad sense) are adequately represented by their Fourier series—that is, over the interval $(-\pi, +\pi)$ the Fourier series of these functions not only converge, but converge to the appropriate values. In essence, Dirichlet showed that Fourier's original claim about trigonometric representability holds true for those functions which Fourier and his contemporaries were likely to recognize as functions. But this was hardly the end of the matter, for many questions remained—among them one which gripped Cantor: Given that a function is adequately represented by a trigonometric series, by how many such series can it be represented? Cantor's uniqueness theorems established that, for certain types of function, the answer is "one." Of particular interest to us is that Cantor's desire to characterize these functions led him to consider complex infinite structures which have since become central to mathematical set theory.

Although it was a tremendous achievement within analysis, Cantor's first uniqueness theorem of April 1870 involved no striking advances toward contemporary set theory. Much the same can be said of Cantor's January 1871 amendment to his first paper—with, as Joseph Dauben points out, one small exception.[28] Cantor initially considered only functions which are represented by a trigonometric series at every point within a given interval. And his first uniqueness theorem covers only these functions. However, he soon turned his attention to functions whose trigonometric representation fails (either through non-convergence or convergence to the wrong value) at certain points (known as "exceptional points"). In his January 1871 note he allows for finitely many exceptional points in finite intervals—thus inviting the obvious question: Could the uniqueness theorem

CANTOR THE ANALYST   29

also be extended to cases involving infinitely many exceptional points in a finite interval? Since a serious attack on this problem called for insight into the topology of the real line, Cantor found himself driven (by the λόγος itself, as Plato would say) to consider infinite structures. Some of the results of these investigations are reported in Cantor's 1872 paper on his successful extension of the uniqueness theorem.

Let me give some details from that paper. According to the Bolzano-Weierstrass theorem, every bounded infinite set of points on the real line has a limit point (a limit point being one whose every neighborhood contains infinitely many members of the set). So, given any bounded infinite point-set (*Punktmenge*) P, the set P^1 composed of the limit points of P is non-empty. If P^1 is infinite, then the set P^2 composed of the limit points of P^1 will also be non-empty. However, if P^1 is finite, then P^2 will be empty, as will P^3 (the set of the limit points of P^2), P^4 (the set of the limit points of P^3), and so on. Whenever P^n is finite and non-empty, Cantor calls the original P a "point-set of the nth kind." This concept turned out to be of great value, for Cantor managed to prove that his uniqueness theorem can be extended to any function whose trigonometric representation over the interval $(0,\pi)$ fails only for the members of a point-set of the nth kind (where n is some natural number).

This is a most impressive result. But what exactly is its set theoretic significance? Contrary to what one might expect, the feature of this paper which was to play the most important role in the development of contemporary set theory was not simply Cantor's recognition that point-sets are important objects of mathematical study. At this stage in his career, Cantor seems to have been partly under the spell of commonsense plurality talk and his notion of "point-set" was rather impoverished. For example, he denies the existence of empty point-sets – just as ordinary language excludes the existence of empty pluralities. For the sake of easy exposition, I indicated that P^m is empty if P is a point-set of the nth kind and $n<m$. But what Cantor actually suggests is that, under these circumstances, P^m does not exist – just as there is no plurality of coins in my pocket if my pocket is empty. A more serious weakness in Cantor's early conception of set is his apparent lack of inclination to carry the process of set formation past the level of point-sets. As Michael Hallett points out, Cantor does treat point-sets as objects to which mathematical operations can be applied.[29] But these operations do not include the formation of sets of point-sets, sets of sets of point-sets, and so on. This is reminiscent of the resistance of ordinary language to the for-

mation of pluralities of pluralities. (On this last topic, see chapter 3.)

A concern with sets is not in itself terribly significant if one's notion of "set" remains crude. Of considerably greater set theoretic importance is the *ordering* Cantor describes in his discussion of the point-sets P^n (the "derived sets," as Cantor calls them). In fact, as Joseph Dauben notes, Cantor himself testifies that it was his reflection on this ordering which compelled him to develop his theory of transfinite ordinals.[30] Let's try to imagine how this might come about. Given the complex structure of the real line, there could very well be non-trivial point-sets which are not of the nth kind for any natural number n. Let P be such a point-set. Then for every n, P^n is non-empty. Furthermore, in some interesting cases, the intersection of all the P^n's will also be non-empty. Let's call this intersection P^ω. Then it is natural to form the set of all the limit points of P^ω – which gives us $P^{\omega+1}$. If P is not of the $(\omega+n)$th kind for any natural number n, then we will be led to form $P^{\omega+\omega}$. And we're off! The way is open to iterate the formation of derived point-sets through more and more remote transfinite ordinals – and the very notion of transfinite ordinal is at hand to be abstracted from this particular context.

We see then that Cantor's theory of transfinite numbers, the heart and soul of contemporary set theory, was not an alien intrusion into the history of mathematics. On the contrary, it grew naturally out of Cantor's very mainstream study of Fourier series – a study which demanded new tools to deal rigorously with the complex structure of the real line. One such tool was an operation (the formation of derived point-sets) whose transfinite iteration seemed not only to be conceivable, but to be positively required for the comprehensive categorization of ill-behaved functions (in Dirichlet's broad, anarchistic sense of "function"). Being at first an unassuming offspring of analysis, only years later did the Cantorian theory of the infinite attain its Oedipal domination of its parent. Nowadays, classical analysis is just one more candidate for reduction to a powerful set theory. And the continuum of real numbers is just one more mathematical array to be located within a comprehensive set theoretic structure. Yet in its infancy, set theory owed not only its (rather limited) mathematical respectability, but even its form and existence to the classical analysis which it was eventually to engulf.

In order to avoid misconceptions (of which there are far too many surrounding Cantor), I should note that the founder of transfinite arithmetic was not entirely blind to set theory's potential as a mathematical adhesive and foundation. By the 1880s, after seriously pondering the ontological status of numbers (both finite and infinite),

Cantor began to appreciate the value of mathematical sets as tools for conceptual analysis and unification. In particular, he realized that the equinumerosity of sets could serve as a criterion for the identity of cardinal numbers, since the cardinal number belonging to a set x is identical to the cardinal number belonging to a set y if and only if x and y are equinumerous. This means that cardinals are abstract objects in the sense discussed in chapter 1.

In fact, Cantor's mathematical sets are themselves abstract objects. (They are not simply commonsense pluralities of uncommonsensical objects.) In 1895, Cantor regarded *structures* as the fundamental mathematical objects – a structure being an aggregate whose identity is determined both by its elements and by the *ordering* of those elements. For example, the structures $<0,1>$ and $<1,0>$ are distinct in spite of being extensionally equivalent (i.e., in spite of having the same elements: 0 and 1). For the ordering of the elements of the first structure is the reverse of that of the second. However, the extensional equivalence of these structures can be taken as a sign of the identity of the corresponding sets. So sets can be taken to stand in the same relation to extensionally equivalent structures that, say, ordinals stand in to isomorphic structures (an analogy which Cantor was perhaps the first to appreciate). Or we might say that we acquire our concept of mathematical set by abstracting from those properties which allow us to distinguish between extensionally equivalent structures. Sets would then be "structureless structures."

Having reached the level of mathematical sets, we are, according to Cantor, only one act of abstraction away from cardinal numbers. Nowadays, in accounting set theoretically for numbers, one normally points out that a certain sequence of sets has the essential structural properties of a given sequence of numbers. This is taken to show that the sets can do all the mathematical work of the numbers. Cantor, on the other hand, stressed the abstractive transformation of thought and discourse about sets into thought and discourse about numbers. Thus, he seems to establish not that set talk can be *substituted* for number talk, but rather that set talk enables us to render number talk meaningful. Cardinal numbers, says Cantor, are "intellectual images or projections of sets" produced by our abstracting from those qualities which allow for discrimination between equinumerous sets.[31] A cardinal number is an abstract object having only those properties shared by certain equinumerous sets. Hearkening back to the Greek notion of ἀριθμός, Cantor notes that his cardinals take on the appearance of sets of "pure units" – these units being objects which have been stripped of everything but their mutual discreteness.

Although this particular approach to cardinals is largely foreign to the mainstream of current set theory, we can nonetheless detect here an early appreciation of set theory as a source of fundamental conceptual tools: Cantor recognizes that we can use set theoretic concepts to provide an analysis of concepts from other areas of mathematics. I should also note that Cantor's rehabilitation of the Greek ἀριθμός notion is not really a subversion of the modern, Newtonian concept of number. As we shall see below, negatives and irrationals remained full-fledged numbers for Cantor.

8. Numbers, Functions, Sets

When, in the course of his study of Fourier series, Cantor contemplated the structure of the real line and was led to introduce the transfinite ordinals, he benefited from two great conceptual achievements: the transformation of the concept "number" and the development of the concept "function."

From the time of Viète on, the acceptance by mathematicians of new species of numbers came to be influenced less and less by the concrete contents of direct experience and more and more by the internal standards and requirements of mathematics and mathematical physics. By the 1880s, Cantor could argue with some success that the introduction of numbers within mathematics is to be governed only by certain standards of consistency, intelligibility, and coherence. This feeling of creative freedom aided the development of set theory not only because it left Cantor uninhibited about introducing transfinite numbers, but also because it helped to supply him with a positive mathematical reason for doing so. Prior to the birth of transfinite arithmetic, Karl Weierstrass, Richard Dedekind, and Cantor himself formulated rigorous theories of arithmetical continua—continua constituted not by some sort of seamless "flow," but by utterly discrete numbers. It was the complex structure of these continua of distinct points which led Cantor to introduce both his infinite ordinals and his infinite cardinals (the latter in connection with his novel proof of Liouville's theorem about the distribution of transcendentals).

Cantor's transfinite arithmetic also owes much to the long unstable notion of function. Oresme and the Oxford Calculators revealed the mathematical tractability of certain relations of functional dependence between continuously changing magnitudes. As mathematical techniques improved, the class of functions which were mathematically treatable grew; and the very concept "function" widened. By Cantor's time, mathematicians were in a position to confront functions so ill-behaved that adequately accounting for them required a detailed grasp of the structure of the real line. Cantor introduced his theory of linear continua because he felt that analysis, the mathematical science of functions, demanded it. Similarly, analysis provided both the inspiration and justification for his introduction of transfinite numbers.

Mathematical set theory owes its distinctive infinitary character not to any natural extension of commonsense set talk, but rather to the very uncommonsensical demands made upon mathematicians by their study of relations of functional dependence between mathematical continua. The gap between mundane pluralities and majestic set theoretic structures is a close relative of the "deep chasm" which Hermann Weyl reveals between everyday experiences of continuity and their mathematical representations.[32] In his effort to grasp the familiar notion of continuity, Cantor felt driven to posit structures whose full treatment eventually called both for the transfinite iteration of certain operations and for the extension beyond the finite of the cardinal numbers. Thus was born the systematic study of the infinite which is the very essence of contemporary set theory.

9. Dedekind Cuts

There is a central feature of contemporary set theory whose history we still need to investigate: namely, the belief that sets are as fit to be themselves members of sets as any other objects. We have already seen that Cantor early on treated sets as objects to which mathematical operations can be applied. But he seems never to have been entirely comfortable with applying the operation of set formation to pluralities of *sets*. In an 1899 letter to Richard Dedekind (the other principal founder of mathematical set theory), Cantor does acknowledge that sets can themselves serve as set elements.[33] Yet he remained reluctant to make extensive use of principles which explicitly guarantee the existence of sets of sets. Perhaps his uneasiness stemmed from his recognition that recklessly positing the existence of sets can lead to contradictions. Or perhaps he was influenced by the inability of ordinary language to accommodate pluralities of pluralities. (On this point, see chapter 3.) In the latter case, commonsense notions of set will indeed have played a minor role in the development of set theory — by briefly *retarding* it.

How then did sets of sets become a mainstay of modern set theory? An important *mathematical* motivation for the explicit and enthusiastic admission of sets of sets was supplied by Dedekind's definition of the real numbers and Ernst Zermelo's proof that every set can be well-ordered.[34] An important *philosophical* motivation was supplied by Gottlob Frege and Bertrand Russell's criticisms of definition by abstraction as it was then practiced.[35] We shall briefly examine the contributions of Dedekind and Frege.

First, some background: A *rational line* λ is a linear array of points whose distances from a fixed origin can be expressed by rational numbers (that is, positive or negative fractions) and which are so arranged that every rational number expresses the distance of some point from the origin. Since the point serving as origin is no distance at all from itself, it will be assigned the rational number 0/1 (or, more briefly, 0). The point lying one half unit to the left of the origin will be assigned the rational number −½. The point lying one half unit to the right of the origin will be assigned the rational number ½. And so on.

A line is said to be *dense* if and only if between any two of its points there lies yet another point. λ is clearly dense—for the associated rational numbers, ordered in the standard way, form a dense array. Speaking somewhat poetically, a line is *continuous* if and only if a "one-dimensional knife" cannot pass through it without encountering a point. The line λ, though dense, is *not* continuous—as the following construction shows.

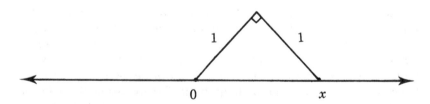

The diagram represents a right isosceles triangle whose base extends from the origin 0 to the position x and whose "legs" are each one unit in length. By the Pythagorean theorem, the distance of x from the origin is the positive square root of 1^2+1^2—that is, it is the positive number whose square is 2. So, if we cut λ at x, our "one-dimensional knife" will encounter a point if and only if there is a rational number whose square is 2. Since it can be shown that there is no such rational number, it follows that λ is not continuous.

The above considerations indicate that a rational line is riddled with "gaps" which correspond to irrational numbers (such as π or $\sqrt{2}$). Note, further, that each gap can be identified by specifying the points which lie to the left (or the right) of it. Alternatively, we could identify gaps by specifying the *rational numbers* assigned to the relevant points. So, for example, the rationals which are either negative or have a square less than 2 could be used to identify the gap x cited above. And since this gap and its ilk correspond to irrational numbers, it follows that certain sets of rationals can be used to characterize irrationals. This insight is the key to Dedekind's definition of the real numbers (that is, the system consisting of both the rational and irrational numbers).

Diverging somewhat from Dedekind's own approach, we shall define a *cut* as a non-empty set C of rational numbers having the following properties: first, a rational number q is in C if and only if every rational number less than q is also in C; but, second, not every rational number is in C. Thus, a cut corresponds to the left hand part of a bisection of the rational line—with the provision that the right hand part of the bisection has no smallest element (although the left hand part might have a greatest one). A cut which has a greatest element is taken to represent that element. For example, $\{q:q \leq \frac{1}{3}\}$ is taken to represent $\frac{1}{3}$. A cut which has no greatest element is taken to represent the "gap" or irrational number which forms the right hand border of its elements. For example, $\{q:q < 0 \text{ or } q^2 < 2\}$ is taken to represent the positive square root of 2.

Perhaps 'represent' is not the best word to use here. My point is simply that Dedekind does not *identify* cuts with real numbers. Rather he uses cuts as tools for fixing the identity of reals. Dedekind is essentially employing a form of definition by abstraction: he uses characteristics of cuts to specify an identity criterion for real numbers. Dedekind says that real numbers "are defined by" or "produce" or "correspond to" cuts. And he notes (or, rather, he would have noted —had he defined cuts in just our way) that the extensional equivalence of cuts is an adequate identity criterion for the associated reals. For the real number corresponding to a cut C is identical to the real number corresponding to a cut C' if and only C and C' have exactly the same elements. Whether Dedekind hereby provides a philosophically adequate grounding for discourse about real numbers will be the topic of our next section. For the moment, I merely note that if (contra Dedekind) cuts *were* identified with real numbers, then sets of reals would become sets of *sets* and the application of set forming operations to pluralities of sets would insinuate itself into the heart of mainstream mathematics.

10. Frege on Abstraction

Gottlob Frege (1848–1925) is one of the two greatest logicians of all time (the other being Aristotle). A sign of his genius is that many of his most brilliant insights now seem almost banal — so thoroughly have his ideas infected the intellectual life of today's logicians. It is tempting to launch into a general review of Frege's accomplishments. But, restraining ourselves courageously, we shall consider only his views on abstraction.

Frege regarded abstraction as an attempt to establish object-hood: he saw it as an attempt to establish that a proper name such as '$\sqrt{2}$' refers to an item whose identity is determined by its possession of certain specifiable properties. Not just any old entity deserves the title of "object" in this special sense. Objects are, first of all, *objective*: unlike the transitory, subjective features of a perceptual experience (features which depend for their very existence on the particular condition of the individual perceiver), *objects* are capable of presenting themselves again and again, in a variety of guises and settings, but always as identically the same. Our identification and reidentification of objects is made possible by their having and retaining characteristic attributes which we capture and publicly express in language. Hence, if we wish to establish that an entity of some sort is an object, we must be prepared to indicate explicitly those abiding characteristics which allow us to identify and reidentify that entity. At a linguistic level, this will involve our specifying the conditions under which certain identity statements are true. For example, we can show that '$\sqrt{2}$' refers to an object only if we can specify the conditions under which sentences of the form

$$\sqrt{2}=x$$

are true (where x is any proper name or definite description in the mathematical language under consideration). So an essential step in establishing the object-hood of $\sqrt{2}$ is the specification of an adequate identity criterion for $\sqrt{2}$. We have seen that Dedekind's use of definition by abstraction supplies us with an identity criterion for real numbers. But does it provide us with an adequate one?

In our rendering of Dedekind's procedure, we noted that an identity statement of the form

[the real number corresponding to the cut C]=
[the real number corresponding to the cut C']

is true if and only C and C' are extensionally equivalent. Thus we can certainly specify the conditions under which at least some identity statements concerning real numbers are true. But how general is our account? Suppose we were confronted with an identity statement of the form

[the real number corresponding to the cut C]=$\sqrt{-1}$.

Does our account allow us to specify the conditions under which formulas of *this* sort are true? Clearly, the above formula is true if and only if $\sqrt{-1}$ is the real number corresponding to a certain cut C' which is extensionally equivalent to C – that is, if and only if there is a cut C' extensionally equivalent to C such that

[the real number corresponding to the cut C']=$\sqrt{-1}$.

We see, then, that we *are* able to specify truth conditions for the above identity statement. But in so doing we have produced a new identity statement whose truth conditions remain unclear. We can, of course, easily specify truth conditions for this new formula: it is true if and only if there is a cut C'' extensionally equivalent to C' such that

[the real number corresponding to the cut C'']=$\sqrt{-1}$.

But this maneuver merely saddles us with yet another identity statement whose truth conditions need to be specified. Indeed, if we carried on in this way, we would generate an unbounded sequence of identity statements whose truth conditions are never fully clarified. The moral is that definition by abstraction (as practiced by Dedekind) has not supplied us with an adequate identity criterion for real number identity statements. For we can easily produce examples whose truth conditions we cannot state without falling into a vicious infinite regress. Thus, Dedekind has not established the object-hood of real numbers.

Frege's solution to difficulties of this sort is to identify would-be objects with items whose objectivity is uncontested and which satisfy the limited conditions for identity laid down in the course of definition by abstraction. For example, in the present case, we have already determined that [the real number corresponding to the cut C]=[the real number corresponding to the cut C'] if and only if C and C' are extensionally equivalent. Our task, then, is to find objects which somehow "correspond" to cuts and whose identity is determined by the extensional equivalence of the cuts to which they correspond. Having found such objects, we shall stipulate that they are the referents of expressions such as '$\sqrt{2}$'. And this will certainly guarantee that $\sqrt{2}$ is an object.

The right sort of objects are (in the case we are now considering) not at all hard to find. Let each cut "correspond" to *itself.* Since cuts are sets, they are identical just in case they have the same members. So the identity of cuts is determined by the extensional equivalence of the cuts to which they "correspond" — that is, the identity of cuts is determined by their very own extensional equivalence. To state the point less circuitously: cuts are perfectly capable of themselves serving as real numbers. Since cuts have all the mathematically important properties of real numbers, there is no mathematically compelling argument against their *being* real numbers. As Russell argues:

> ... we find that [cuts] fulfil all the requirements laid down in Cantor's definition [of the real numbers], and also those derived from the principle of abstraction. Hence there is no logical ground for distinguishing [cuts] from real numbers. If they are to be distinguished, it must be in virtue of some immediate intuition, or of some wholly new axiom, such as, that all series of rationals must have a limit. But this would be fatal to the uniform development of Arithmetic and Analysis ... and would be wholly contrary to the spirit of those who have invented the arithmetical theory of irrationals. [The identification of cuts and real numbers] removes what seems, mathematically, a wholly unnecessary complication, since, if [cuts] will do all that is required of irrationals, it seems superfluous to introduce a new parallel series with precisely the same mathematical properties.[36]

Since Dedekind's scrupulous discrimination between cuts and reals was found not to be mathematically defensible, it was abandoned. Real numbers became cuts.

But cuts are sets. And real numbers are objects whose membership in sets is essential to mathematics as we know it. Thus, the notion that sets of sets are mathematically respectable and, indeed, mathematically indispensable objects flowed fairly directly from Dedekind's effort to clarify the system of real numbers — an effort which was motivated by purely *mathematical* concerns.[37] Dedekind was indeed influenced by *philosophical* views about the proper way to render a mathematical vocabulary meaningful. But these views served only to *retard* the acceptance of sets of sets. Thus, Frege's attacks on definition by abstraction (as practiced by Dedekind) served primarily to remove philosophical impediments to the mathematically natural development of mathematical set theory.

As I indicated earlier, our review of set theory's historical roots would be valuable if it did no more than remind us that set theory *has* a history – a history which philosophers can ignore only at their own peril. We have, however, accomplished even more. For we have confirmed that the first of the myths cited above is historically false. Set theory does not owe its essential characteristics to any intrusion of commonsense notions into the mathematical domain. Set theory's most central and distinctive features (the extension of arithmetic into the transfinite and a commitment to sets of sets) sprang from mathematical responses to mathematical problems – problems arising primarily within the theory of functions and real numbers.

Having seen that commonsense notions of set *did not* significantly influence the development of set theory, we shall (in the following chapter) discover that they *could not* have. We shall see that there is a major lack of fit between mundane sets and their mathematical namesakes. And we shall thus render even more implausible the second of the myths mentioned above (Myth #2 having already been undermined to some extent by our debunking of Myth #1).

III
COMMONSENSE SETS

> From the very beginning ... the teacher should see to it that students acquire, by their own effort, an understanding of the concept of 'set', building largely upon examples that they have encountered in their social life, their experiences at school and in the world about them. – Office for European Economic Coordination, *Synopses for Modern School Mathematics*[38]

1. Nausea

The March 25, 1974 issue of the German periodical *Der Spiegel* features a cover photograph of a world-weary youngster under the blaring headline, "Macht Mengenlehre krank?" ("Is set theory making our children sick?"). I doubt that set theory itself has caused many physical indispositions. But an educator could well induce an epidemic of vertigo by following the OEEC directive cited above. The OEEC *Synopses* assume that children (and, presumably, adults too) can reasonably be expected to grasp the concept of mathematical set "by their own effort" *prior* to learning anything about mathematical set theory. Our previous chapter suggests that this is a mistake. Since commonsense notions of set played no significant role in the development of mathematical set theory's most distinctive features, it seems most unlikely that one could acquire an adequate concept of mathematical set merely by marshalling such notions.

Yet doubts may remain: perhaps, for some historically obscure reason, commonsense discourse (or "ordinary language," as philosophers call it) contains mechanisms for producing an essentially sound notion of mathematical set – mechanisms which are effective even in the absence of a thorough grounding in mathematical set theory. In the following sections, we shall examine the most promising attempt to show that such mechanisms exist. Since this will require that we venture into the domain of ordinary language analysis, let me first express some general reservations about appealing to the authority of ordinary usage to answer philosophical questions. I shall then try to explain why these reservations need not prevent us from carrying out the brief inquiry into ordinary language which follows.

2. The Court of Ordinary Language

Much philosophical work in this century has been based on the authority of our ordinary use of language. In some extreme cases, this move is a manifestation of the view that ordinary usage is the highest court of appeals in matters philosophical and that every bearer of linguistic intuitions is competent to rule on philosophical questions. A more moderate view is that an appeal to ordinary usage can serve merely to shift the burden of proof onto philosophers whose theories clash with that usage. This would mean that ordinary language is to be consulted not as a way of authoritatively settling philosophical disputes, but rather as a way of clarifying what should count as such a settlement.

My own feeling (and, at the moment, it is little more than a feeling) is that any disputes, whether in philosophy or physics, which are meant to be settled in a rational way can be, often are, and generally should be regulated by norms of rationality which are neither human universals nor idiosyncracies of particular pretheoretic language forms but, instead, are products of a more or less self-conscious effort to produce rationally validated theories while, simultaneously, determining what is to count as rational validation. I don't claim to be at all clear about what these standards of rationality (which we might call "meta-theoretical norms") are at the moment. Still, the usual lists of theoretical virtues (generality, simplicity, consistency, clarity, etc.) do seem to capture certain aspects of these norms. And in certain cases, we can indicate (albeit vaguely) what rational behavior amounts to. For example, it seems safe to say that philosophical logic should, among other things, be in the business of replacing the vague and possibly incoherent notions which we often find expressed in ordinary discourse with determinate and coherent concepts. That is, someone engaged in philosophical logic should not necessarily look to ordinary language either as a source of rational validation or even as an arbiter of what is to count as such validation. Instead, the philosophical logician's evaluation of a theory should be regulated by meta-theoretical norms which are neither rigid laws written on our genes nor cultural artifacts embedded in ordinary discourse, but rather reflectively constructed expressions of our desire to be rational and our current understanding of what that desire amounts to.

Does this mean that ordinary language analysis has no place in philosophy? Should we ignore the work of philosophers whose responses to questions about the concept "set" have included analyses of our everyday set talk? Certainly not. It is surely possible to clarify a technical concept by tracing it back to its roots in everyday discourse (*provided that it has such roots*). This is not to suppose that ordinary language protoconcepts are generally clearer or in other ways preferable to their offspring in the various specialized theoretical disciplines. It is merely to acknowledge that in mastering or even inventing the patterns of linguistic usage which render a technical terminology meaningful, it is often helpful to fall back on nontechnical linguistic skills – skills which may be sufficiently basic to be considered part of ordinary language.

By thus bowing in the direction of everyday discourse, we in no way acquiesce to the notion that technical theories must conform to any sort of commonsense worldview. For even though technical language can be *clarified* through some sort of correlation with ordinary language, it does not follow that our *evaluation* of a theory which has been thus clarified should involve any appeal to commonsense notions. We see, then, that employing ordinary language analysis for the purpose of clarification is much different from employing it for the purpose of validation. Since I have questioned the latter move but not the former, my remarks about the alleged authority of ordinary language should not keep us from exploring everyday set talk in an effort to clarify mathematical set theories. In fact, we shall now engage in just such an exploration. As I have already indicated, we shall find that a mere grasp of commonsense notions of set will *not* supply us with an even remotely adequate appreciation for mathematical sets. In fact, I doubt whether *any* "commonsensical" clarification of mathematical set theories is either possible or needed. My point is simply that we can be sure of this only *after* an investigation of ordinary language. And such an investigation is warranted even if its results are wholly negative.

3. Black on Sets as Commonsense Objects

In his essay "The Elusiveness of Sets," Max Black argues that our mastery of the familiar device of plural reference is sufficient to supply us with a fundamentally reliable grasp of mathematical sets. To express the point somewhat differently: Black argues that commonsense sets (construed in a certain way) are objects of the same type as mathematical sets. As a convenient shorthand, we might say simply that Black "identifies mathematical and commonsense sets" or "claims that mathematical sets are commonsense objects."

Black occasionally seems to suggest that a technical terminology (such as mathematical set talk) can be rendered meaningful only through its being connected up with preexisting ordinary language skills. But it would be uncharitable to saddle Black with such a dubious position – a position which appears no more plausible than the more general claim that it is impossible to master any linguistically isolated system of novel linguistic skills (that it is impossible to teach someone a language game without revealing crucial similarities between the moves in that game and moves in games which are already familiar – so that, say, a child's *initial* understanding of language must be supported by a *prior* grasp of determinate language games). I gather that Black is actually making the much less dubious claim that one can plausibly assert both that sets are commonsense objects and that they are apparently uncommonsensical "abstract entities" only when one has shown how our talk about those "abstract entities" can be connected up with ordinary language. If this connection cannot be made, then sets may indeed be "abstract entities," but they are not commonsense objects.

Black spends much of his essay trying to show that various philosophical interpretations of set theory either fail to account for certain set theoretic truths or fail to be intelligible from a commonsense point of view. For example, in order to discredit the view that a set comes into being through a human mental act of "collection" or "unification" directed at its members, Black reminds us that mathematical set theories posit the existence of various sorts of sets without any regard to whether the members of those sets are, ever have been, or ever will be objects of thought. Indeed, set theorists *must* do this if they wish to remain philosophically respectable while retaining their vision of an enormously large and complex set theoretic universe whose members are, for the most part, themselves enormously large and complex. For it would be quite outlandish to posit the existence of as many human mental acts of unification as there are sets in the universe of, say, Zermelo-Fraenkel set theory (ZF). And it seems just as outlandish to suppose that any individual human mental act could be sufficiently complex to count as the "unification" of the members of any of the larger and more complex sets available in the universe of ZF. So this business of "mental collection" is an example of a philosophical interpretation which fails to account for certain presumed truths of set theory.

Black also regards the notion of "mental collection" as a piece of philosophical speculation which cannot be explicated in commonsense terms and which, therefore, cannot be used to establish that sets are commonsense objects. Black's "argument" here is basically just a declaration, repeated several times, that the proponents of "set formation through mental unification" have not managed to make this notion intelligible to him. And perhaps this is all the "argument" Black need offer here. For the burden of proof would seem to lie with those (if there really are any) who claim that "mental unification" talk has some correlate in ordinary language. I should note, however, that Black also seems to think that mental unification talk is nonsensical in some absolute way (that it is a mere *flatus vocis*). And insofar as he means to make such a strong and implausible claim, the burden of proof shifts to him.

In spite of occasional lapses into ordinary language dogmatism, Black's discussion of "mental collection" is generally reasonable and convincing. I find myself less favorably disposed toward Black's discussion of abstraction in the section entitled "Sets as Arising from Equivalence Relations Between Properties." Black argues that we cannot use abstraction to account for any "abstract entity" talk unless we presuppose a certain amount of set theory. So we cannot use abstraction to account for set theory itself. This argument appears satisfying only if we follow Black in adopting an overly narrow view of abstraction. Still, Black's point is fundamentally correct — for abstraction, even when properly construed, does not supply us with any commonsense account of mathematical sets. I shall postpone further discussion of this point until our next chapter.

4. Sets and Plural Reference

Of course, Black's project is not entirely negative. His criticisms of unsuccessful accounts of the concept "set" are merely a prelude to his own positive effort at explicating this concept. Black shows that our everyday ability to refer to several objects at once places us within reach of "the abstract notion of a set as *a number of things considered together . . .* or *several things referred to at once.*"[39] And he argues that mathematical sets can be regarded as pluralities of precisely this sort.

Plural reference is an entirely familiar linguistic phenomenon whose philosophical significance has been only rarely appreciated. As Black points out:

> The most obvious ways of referring to a single thing are by using a name or a definite description: 'Aristotle' or 'the president of the United States'. Equally familiar, although strangely overlooked by logicians and philosophers, are devices for referring to several things *together:* 'Berkeley and Hume' or 'the brothers of Napoleon'. Here, *lists* of names (usually, but not necessarily, coupled by occurrences of 'and') and what might be called "plural descriptions" (phrases of the form 'the-so-and-so's' in certain uses) play something like the same role that names and singular descriptions do. Just as 'Nixon' identifies *one* man for attention in the context of some statement, the list 'Johnson and Kennedy' identifies two men at once, in a context in which something is considered that involves both of them at once. And just as 'the President of the United States' succeeds in identifying one man by description, so the phrase 'the American presidents since Lincoln' succeeds in identifying several, in a way that allows something to be said that involves all of them at once.[40]

Peter Simons provides a somewhat fuller typology of plurally referring expressions:

Plural proper name	Benelux
Plural definite description	the fishermen of England
Plural demonstrative	these books
Plural personal pronoun	they
Name list	Tom, Dick, and Harry
Mixed term list	Jason and the Argonauts[41]

As Black emphasizes: "The notion of 'plural' or simultaneous reference to several things at once is really not at all mysterious. Just as I can point to a single thing, I can point to two things at once – using two hands, if necessary; pointing to two things at once need be no more perplexing than touching two things at once."[42]

Having reviewed the ill-appreciated device of plural reference, Black proceeds to make a claim to which I take no exception – namely, that at least some commonsense set talk is a fundamentally equivalent alternative to forms of discourse involving plural reference:

> One primitive use of the word 'set' is as a stand-in for plural referring expressions ... If I say "A certain set of men are running for office" and am asked to be more specific, then I might say, "To wit, Tom, Dick, and Harry" – or, in the absence of knowledge of their names, I might abide by my original assertion. One might therefore regard the *word* 'set', in its most basic use, as an indefinite surrogate for lists and plural

descriptions. To know how to use the word 'set' correctly at this level is just to know the linguistic connections between such uses of 'set' and the uses of more definite multiply-referring devices.[43]

Similar remarks apply to other commonsense set terms (beyond the word 'set' itself). Consider, for example, the sentences:

> Your collection of books includes all of my favorites.
> All of my favorites are among the books you have collected.

The second of these sentences can plausibly be regarded as a faithful rephrasing of the first — even though the second replaces the singular expression 'your collection of books' with the plural expression 'the books you have collected'.

So far so good. From this point on, however, Black's claims become more and more suspect. We saw in our last chapter that a distinctive feature of mathematical sets is their ability to have other mathematical sets as members. So, if one is to show that mathematical and mundane sets are objects of the same sort, one must show that there are such things as mundane sets of sets. With this end in mind, Black argues essentially as follows. When one refers to a set, one is referring to several objects at once (the so-called *members* of the set). For example, when I refer to your collection of books, I am referring to the books you have collected. When I refer to your social circle, I am referring to the people with whom you socialize. Briefly, then, sets are pluralities. So, to refer to a set of sets would be to refer to several *pluralities* at once. Is this something which ordinary English is capable of doing?

Black believes that it is. He notes that lists of terms can function as plurally referring expressions and that such lists can include terms which themselves function as plurally referring expressions. Further, he suggests that this concerted use of plural expressions allows us to refer plurally to pluralities. For example, the term 'the Montagues' refers plurally to certain people — as does the term 'the Capulets'. So the term list 'the Montagues and the Capulets' must refer plurally to two pluralities of people — that is, the Montagues and the Capulets must be a plurality consisting of two pluralities. We have supposed that when we refer to a set of sets, we are referring to the corresponding plurality of pluralities. So, were we to refer to "the set consisting of the set of the Montagues and the set of the Capulets," we would really just be referring to the Montagues and the Capulets.

Or would we? Let M be the set of the Montagues, let C be the set of the Capulets, and let {M,C} be the set whose only members are M and C. If '{M,C}' (or rather its unabbreviated counterpart) is a commonsense "surrogate" for 'the Montagues and the Capulets', then it should be clear to common sense how statements involving the former locution are to be rephrased as statements involving the latter. I maintain that this is decidedly unclear.

Suppose that there are 127 Montagues and 133 Capulets. Then the Montagues and the Capulets are 260 in number. Now consider the following assertions:

A. M has 127 members.
B. {M,C} has 2 members.

We can easily rephrase A as a statement involving plural reference. 'M' is taken to be a surrogate for 'the Montagues'; so we replace the former expression by the latter. And we replace 'has 127 members' by 'are 127 in number'. This yields:

A'. The Montagues are 127 in number.

No problem here. Now let us apply the same procedure to B. '{M,C}' is taken to be a surrogate for 'the Montagues and the Capulets'; so we replace the former expression by the latter. And we replace 'has 2 members' by 'are two in number'. This yields:

B'. The Montagues and the Capulets are 2 in number.

Of course, the problem here is that B is true, whereas B' is false (the Montagues and the Capulets being, in fact, 260 in number). So B' is certainly not an accurate rephrasing of B.

The source of our difficulty is that the expression 'the Montagues and the Capulets' is naturally taken to refer to all the Montagues and Capulets *regarded as members of a single plurality*. So this expression is not a reliable stand-in for the set-of-sets expressions to which it allegedly corresponds. (We might express the point thus: If a set is to be a member of a set, it must be a single thing rather than a plurality of things. For were it a plurality, it would not be *a* member – it would be several.) The bare fact that plural expressions can figure within larger plural expressions fails to imply that there is a generally trustworthy technique for translating sets-of-sets talk into the commonsense language of plural reference. In fact, I'm confident that there simply is no such technique.

I should note, however, that Black has an argument up his sleeve which I have not yet mentioned. Black maintains that "ostensibly singular" expressions such as 'the Montague family' are surrogates for plural expressions such as 'the members of the Montague family'. Thus, for example, we can take

> The Montague family resided in Verona

to mean that the *members* of this family resided there. We are to suppose, then, that 'the Montague family' and kindred expressions, "... look superficially like singular descriptions but really serve to refer to several things at once."[44] So 'the Montague family' names a plurality just as 'the Montagues' does (and, in fact, these two expressions name the very same plurality).

Now consider the plural definite description

> the families who figure prominently in *Romeo and Juliet*.

This expression refers plurally to the Montague and Capulet families – which, as we have just decided, are themselves pluralities. So this expression seems to name a plurality of pluralities. Note that here there is no fusion of Montagues and Capulets. For the families who figure prominently in *Romeo and Juliet* are 2 (not 260) in number. We seem, then, to be referring successfully to the Montagues and the Capulets *regarded as distinct pluralities*. And given such pluralities of pluralities, we should be able to make sense of sets of sets in terms of plural reference.

But wait just a minute! Something has gone seriously wrong here. Consider again the three expressions:

α. the families who figure prominently in *Romeo and Juliet*.
β. the Montague family and the Capulet family.
γ. the Montagues and the Capulets.

α and β are taken to refer to precisely the same things. So, since the items named by α are 2 in number, the items named by β must also be 2 in number. Furthermore, 'the Montague family' and 'the Capulet family' are taken to refer to precisely the same items as 'the Montagues' and 'the Capulets'. (That's the point of saying that 'the Montague family' is an *ostensibly* singular expression.) So β must refer to the same items as γ (since the items listed in β are the same as those listed in γ). It follows that the items named by γ are 2 in number. Inasmuch as these items are (we have supposed) 260 in number, we may conclude that 2=260.

Of course, 2≠260. So at least one of our premises must be false. But which one(s)? Surely, α and β do refer to the same items. And, surely, these items are 2 in number. But this obviously means that 'the Montague family' and 'the Capulet family' each refer to *one* item.

So 'the Montague family' and 'the Capulet family' are *genuinely*, not just ostensibly, singular expressions. And it is false that β and γ refer to the same items (for γ is a conjunction of uncontroversially plural expressions). It is equally false that α names a plurality of *pluralities*. Black's initially promising argument is a failure. And we continue to regard it as most dubious that the notion of plural reference can supply us with a uniform analysis of sets-of-sets talk.

But perhaps we have no need of an account which appeals to plural reference. Don't genuinely singular set expressions such as 'family' and 'group' allow us to speak perfectly intelligibly about commonsense sets of sets? For example, there is nothing at all mysterious about referring to *a group of families*. Do we not thereby refer to a set of sets of the same basic type as a mathematical set of sets? Let us see.

* * *

Friar Laurence: See that group of families over there?

Friar John: You mean the one with Benvolio in it?

Friar Laurence: Are you crazy? Benvolio isn't a family!

Friar John: But look! He's in that group you were just pointing to.

Friar Laurence: Oh, you mean the group of *people*. I was talking about the group of *families*. You know, the Montagues and the Capulets.

Friar John: But Benvolio *is* a Montague.

Friar Laurence: Yes, yes. But that just means he's a member *of a member* of the group I'm talking about.

Friar John: Well alright, if you insist. What exactly did you want to say about this "group"?

Friar Laurence: It's very small.

Friar John: [pointing to a crowd of 260 people] *That* group over *there* is very small?!?!

Friar Laurence: Of course! You just admitted that it has only two members.

Friar John: Let me see if I've got this straight. All you're trying to tell me is that there are two families over there, right?

Friar Laurence: Well, not exactly. I was really saying that there is a two-membered *group* of families over there.

Friar John: You're talking about this *funny* group of yours, the one that Benvolio *doesn't* belong to.

Friar Laurence: Yes. He *is* a member of its union though.

Friar John: I don't want to hear about it!

* * *

Unbeknownst to Friar John, Friar Laurence has invented mathematical set theory. Confusion arises because Friar Laurence chooses to cloak his pathbreaking innovation in the ill-fitting mantle of commonsense set talk. As George Bealer has emphasized, a crucial difference between mathematical sets and commonsense collectivities is that the latter, but not the former, are presumed to be transitive with respect to membership. (This property is technically known as "$\in$-transitivity" – '$\in$' being the standard symbol for set membership.) Thus, if x and y are mundane collectivities and y is in x, then it is presumed that everything in y is also in x. For example, if the Montague family is in a certain group, then there is a strong presumption that Benvolio, being a Montague, is also a member of that group. (If the Montague family belongs to the local YMCA, then Benvolio, as an individual Montague, has all the rights and privileges of any other member.) In the case of mathematical sets, on the other hand, there is no presumption in favor of $\in$-transitivity. There are indeed an incomprehensibly large number of $\in$-transitive sets, but there are just as many that lack this property.

Mathematical sets and mundane collectivities also differ with respect to their identity conditions. The identity of a mathematical set is determined entirely by its membership. Commonsense collectivities, on the other hand, can change members while retaining their identity. The death of Romeo did not spell the end for the Montague family. Nor would the birth of a new Montague signal the creation of a new family. Families do pass into and out of existence, but the addition or loss of a member does not in itself guarantee such a transformation. Conversely, as we shall see in section 5, mundane collectivities can be distinct even when they have exactly the same members. No such option is available to mathematical sets.

The moral of this story is that while singular set expressions in ordinary English do indeed allow us to speak intelligibly about collectivities of collectivities, there is an enormous gap between a mastery of these locutions and a grasp of mathematical set talk.[45] As we have seen, mundane plurality talk fares no better (thanks to the resistance of ordinary English to the formation of pluralities of pluralities). We conclude that near cousins of mathematical set talk are to be found, at best, only at the dim periphery of ordinary English.

Our severely compressed critique of some of Black's theses should not blind us to the overall importance of his contribution to the philosophy of set theory. His insistence on the philosophical significance of plural reference and his demand for an intelligible account of mathematical sets has inspired some important work (an example of which we discuss in chapter 9). Black deserves our thanks for directing attention to issues which have suffered far too much neglect.

5. Zermelian Sets

Thus far, my readers have had to depend entirely on my vague indications of the characteristics mathematical sets do or do not have. Yet I have been arguing that an adequate appreciation of these sets is not to be expected from anyone unacquainted with mathematical set theories. It is high time, then, that we examine at least one such theory. There are two good reasons for choosing Ernst Zermelo's pathbreaking axiom system of 1908.[46] First, Zermelo's theory (known, more briefly, as "Z") consists entirely of postulates which are currently regarded as uncontroversial and, indeed, indispensable. Granted, Zermelo's Axiom of Choice was not so regarded when first introduced; but it has since become a core component of set theory. So, if Z turns out to be "uncommonsensical," this won't be because it includes axioms which mathematicians regard as bizarre (that is, the departure from common sense will not be a symptom merely of mathematical heterodoxy). Second, should any significant connection exist between mathematical and mundane sets, one would expect it to be most evident in the earliest efforts at axiomatization—efforts carried out by thinkers who had not yet fully developed a technical vocabulary and an arsenal of technical concepts and skills. If an early effort of this sort turns out to be "uncommonsensical," we can expect the more and more technical transformations it subsequently undergoes to make it only more so. Let us, then, examine Zermelo's axioms and determine whether they match the picture of commonsense sets which we have sketched above.

Empty Set Axiom: There is, in Zermelo's words, a "fictitious" set ∅ which has no members at all.

I gather that, in calling the empty or null set "fictitious," Zermelo is himself noting a departure from our ordinary notion of set. A pack of wolves ceases to exist if all its members die out. A plurality of nothing is no plurality at all. (More on this point in our discussion of the next axiom.)

Pairing Axiom: Given any objects x and y (either or both of which may be sets), there is a set $\{x,y\}$ whose members are exactly x and y.

In those cases where x is not identical to y and neither x nor y is a set, the pairing axiom seems consistent with mundane plurality talk. Given any two non-pluralities x and y, there is a plurality consisting of just x and y. As Peter Simons expresses it: "... whoever admits the existence of at least two individuals admits that of at least one plurality ... one cannot affirm at least two individuals but deny pluralities, for the plurality of two objects just is *them*."[47] A plurality of two individuals simply is those individuals considered plurally. (That's the fundamental point of linking pluralities to the device of plural reference.) On the other hand, if x or y are commonsense pluralities or collectivities the pairing axiom runs up against either the non-existence of pluralities of pluralities or our commonsense resistance to the formation of collectivities which fail to be $\in$-transitive. (Note that the pair set {Montague family, Capulet family} guaranteed by the pairing axiom does not feature Benvolio as a member.) Furthermore, if x is identical to y, the pairing axiom runs afoul of the resistance of ordinary English to the formation of unit sets (sets with just a single member). As Black remarks: "... any transition from colloquial set talk to the idealised and sophisticated notion of ... a 'null set' and of a 'unit set' (regarded as distinct from its sole member) will cause trouble [for someone who regards mathematical sets as commonsense objects]. From the standpoint of ordinary usage, such sets can hardly be regarded as anything other than convenient fictions ... useful for rounding off and simplifying a mathematical set theory. But they represent a significant extension of ordinary use; and nothing but muddle will result from ignoring this point."[48]

Separation Axiom: Given any set y and property P, there is a set $\{x \in y : Px\}$ whose members are exactly those members of y which have P.

Roughly translated into the language of pluralities, this axiom makes the bland assertion that those members of a plurality which share a certain property themselves form a plurality. For example, the Capulets who are left-handed form a plurality – namely, the plurality of left-handed Capulets. So far so good. But note that the property P might be had either by nothing (e.g., let P be the property of non-self-identity) or by just one thing (e.g., let P be the property of having authored *War and Peace*). In the case of such a P, we run into the problems discussed in connection with the previous two axioms. Furthermore, the separation axiom jars with our commonsense conception of singular collectivities (groups, families) because it permits violations of ∈-transitivity. (The ∈-transitivity of y does not guarantee the ∈-transitivity of $\{x \in y : Px\}$.)

Power Set Axiom: Given any set x, there is a set $\mathcal{P}(x)$ whose members are exactly the subsets of x (a subset of x being a set all of whose members are members of x).

Union Axiom: Given any set x, there is a set $\cup x$ whose members are exactly the members of members of x.

Axiom of Choice: Given any set x of non-empty and mutually disjoint sets, $\cup x$ contains a subset among whose members there occurs exactly one member of each member of x.

In all three of these axioms, we once again encounter the problem of iterated set formation (the problem of sets whose members are themselves sets). The power set axiom directly postulates sets of sets. The union axiom yields the empty set when applied to anything other than a set of sets. And the choice axiom can be applied only to sets of sets. So none of these axioms can be regarded as (nontrivially) accurate depictions of mundane pluralities. The axioms fare a bit better when it comes to singular set locutions ('family', 'group', etc.). If x is ∈-transitive, then so is $\mathcal{P}(x)$. So the power set axiom does not introduce any failure of ∈-transitivity and, hence, is at least *consistent* with our commonsense conceptions. (It's far from clear, however, that this axiom is *implied* by those conceptions.) The union axiom also demands no failure of ∈-transitivity (since $\cup x$ is ∈-transitive whenever x is). But if we restrict ourselves to ∈-transitive sets, the union axiom is superfluous (since $\cup x$ is then a subset of x and, hence, is guaranteed to exist by the separation axiom). Since the axiom of choice does not preserve ∈-transitivity, it turns out to be inconsistent with our everyday notion of collectivity.

Axiom of Infinity: There is a set x with the following properties: first, the empty set $\emptyset$ belongs to x; and, second, if an object y belongs to x, then so does $\{y\}$. (So x has among its members the empty set $\emptyset$, the set $\{\emptyset\}$, the set $\{\{\emptyset\}\}$, the set $\{\{\{\emptyset\}\}\}$, and so on.)

The infinity axiom guarantees that there is a set which has infinitely many sets as members. (The pairing and empty set axioms guarantee that there are infinitely many sets. But they do not guarantee that there is a set with infinitely many members.) Here again we run afoul of the empty set, unit sets, and sets of sets. Furthermore, it is unclear that any worldview which deserves to be considered "commonsensical" features a commitment to infinitely many sets — much less a set whose membership features infinitely many sets.

Axiom of Extensionality: If sets x and y have exactly the same members, then x is identical to y.

As Peter Simons emphasizes, commonsense pluralities do obey the extensionality axiom: pluralities which consist of exactly the same objects are identical.[49] Consider, for example, the plural identity

> Matthew, Mark, and Luke are the authors of the synoptic Gospels

which we might abbreviate as

> [Matthew, Mark, and Luke]=[the authors of the synoptic Gospels].

There are essentially only two ways for this assertion to be false. It would be false if either Matthew, Mark, or Luke failed to be among the authors of the synoptic gospels. And it would also be false if one of the authors of the synoptic gospels was neither Matthew, Mark, nor Luke. So if every member of the one plurality is also a member of the other, the assertion that the pluralities are identical is true. We see, then, that Extensionality does accurately state an identity criterion for commonsense pluralities. And it seems likely that, in this case, the formulation of a mathematical axiom was inspired by a commonsense principle. Still, let me express two reservations. First, from a purely mathematical viewpoint, Extensionality is the most dispensable of Zermelo's axioms. As I and others (most notably, George Bealer, Frederic B. Fitch, Nicolas D. Goodman, Michael Jubien, Norman M. Martin, and Rolf Schock) have shown, mathematicians would not be seriously incommoded by the absence of Extensionality – as long as they were supplied with any one of a number of suitable substitutes.[50] We might say, then, that the major contribution which common sense makes to Z comes at a relatively unimportant point. Second, there are many commonsense collectivities which, when treated as *single* objects, are *not* accurately characterized by Extensionality. Peter Simons remarks that, "... in the days of the Empire, three of the orchestras of Vienna had the same personnel: when they played in the Court Chapel they were the Orchestra of the Court Chapel, when they played in the pit at the opera they were the Court Opera Orchestra, and when they played symphony concerts in the Musikverein they were the Vienna Philharmonic. Similarly two committees may have exactly the same members, yet not be one committee."[51] Though collectivities such as orchestras and committees are often cited as commonsensical examples of sets, they do not obey Extensionality: they can be distinct in spite of having the same members. (And we earlier saw that they can be identical in spite of having different members.) Given the mixed bag of set-like objects populating the everyday world, we might say that Extensionality is as much a departure from ordinary usage as it is a reflection of it.

According to the second of the myths we have been examining, one can reasonably be expected to have an accurate conception of mathematical sets prior to learning anything about mathematical set theory. I hope that this contention has been made to appear highly questionable. The features shared by mathematical sets and mundane pluralities or collectivities are few in number and are all of limited mathematical significance. So there is no reason to think that a mastery of everyday set or plurality talk will in itself supply one with anything other than a *mis*conception of mathematical sets. Furthermore, no other sector of ordinary English seems likely to give any more satisfactory results. So let us not demand that toddlers acquire a grasp of mathematical sets "by their own effort." That would indeed be a recipe for making our children sick.

We have devoted enough time to myth-busting. For the remainder of this book, we shall consider views of set theory which stand a better chance of being defensible.

IV
LOGICAL INTERLUDE: INTERPRETABILITY

1. The Age of Logic?

What branch of science (formal or empircal) has made the greatest strides in the last hundred years or so? Theoretical physics comes immediately to mind. And other scientists, such as cell biologists, could point to the unusual vitality of their disciplines. Could any consensus on this ill-defined question be reached among scientifically literate people? Perhaps so, if the advances in *logic* could be made intelligible to those outside the field. For, at least from the biased point of view of a logician, it seems hard to imagine that any other discipline could match the progress made by logic (in the broad sense, including set, model, and recursion theory) since the publication of Gottlob Frege's *Begriffsschrift* in 1879.

These are thoughts well calculated to fill a logician with missionary zeal. Whether or not we are living in the Age of Logic, the important and intrinsically interesting results of the post-Frege era should be made accessible to non-logicians. Of course, accomplishing this task can hardly be the aim of the thin volume you are now reading. The job can probably be done both thoroughly and well only by a sizable stack of books—many of which have already been produced by authors better suited to the task than I am. My only contribution to this endeavor will be to explain (briefly) those technical concepts and results which will arise in our philosophical discussion of set theory. I hope this will not prove too tedious for the logically initiated reader. Remember: even the sophisticated can chance upon some insights while grubbing about in the roots of their discipline. In order to prepare ourselves for our next chapter and to tie up some loose ends from the last, we must now concentrate on the logical theory of interpretability.

2. Interpretation Functions

We owe much of our knowledge of interpretability (as a logical notion) to Alfred Tarski.[52] In the most general sense, a mathematical theory Σ is interpretable in another theory Σ' if there is a function which maps the theorems of Σ into theorems of Σ'. Not only is this the most *general* sense of interpretability – it is also the most vacuous and the least interesting. For suppose that ψ is a theorem of Σ'. Then, given any theory Σ, we can easily define a function i such that $i(\varphi) = \psi$ whenever φ is a theorem of Σ. (Just let i map all the formulas of Σ into ψ, for example.) So, if we are able to identify a single theorem of Σ', then, in this utterly empty sense, any system Σ is interpretable in Σ'.

Clearly, we need to place further restrictions on the interpretation function i if the notion of interpretability is to have any bite. A very minimal restriction is to require that $i(\varphi)$ be a contradiction whenever φ is a contradiction. Then, given that Σ is interpretable in Σ', we could infer that Σ is consistent if Σ' is. For if Σ were inconsistent, then some contradiction φ would be a theorem of Σ. But then $i(\varphi)$ would be a contradiction. So some contradiction would be a theorem of Σ' (since $i(\varphi)$ is a theorem of Σ' whenever φ is a theorem of Σ) and, thus, Σ' would be inconsistent. From the inconsistency of Σ we could infer the inconsistency of Σ'. So the consistency of Σ' would imply the consistency of Σ.

As I said, this is a very weak restriction on the function i. It fails to capture our commonsense notion that a good interpretation or translation of a statement will "explicate the meaning of" or "say essentially the same thing as" the statement being interpreted or translated. One way of approximating the notion of "saying essentially the same thing" is to require that interpretation functions preserve basic logical form. (This would also guarantee that contradictions will be mapped into contradictions and, therefore, that interpretability will imply relative consistency.) In order to make sense of the notion of logical form, it is useful to have a vocabulary of formal logical symbols. I supply the following translation schemes for readers who have not been initiated into the mysteries of symbolic logic. Here 'φ' and 'ψ' stand for statements and 'α' stands for a variable (as discussed in chapter 7, §1). Read

$\neg \varphi$ as   It is not the case that φ
$(\varphi \rightarrow \psi)$ as   If φ, then ψ
$(\varphi \leftrightarrow \psi)$ as φ if and only if ψ

58 LOGICAL INTERLUDE: INTERPRETABILITY

$(\varphi \wedge \psi)$ as φ and ψ
$(\varphi \vee \psi)$ as φ or ψ
$\forall \alpha \varphi$ as   For every α, φ
$\exists \alpha \varphi$ as   There is an α such that φ.

In the case of the symbols '¬', '→', '↔', '∧', and '∨' (which are called "sentential connectives"), it is easy to say what should count as a preservation of logical form. Negations should be mapped into negations, conditionals into conditionals, biconditionals into biconditionals, conjunctions into conjunctions, and disjunctions into disjunctions. More precisely, if i is an interpretation function,

$i(\neg \varphi)$ should be $\neg i(\varphi)$
$i(\varphi \rightarrow \psi)$ should be $i(\varphi) \rightarrow i(\psi)$
$i(\varphi \leftrightarrow \psi)$ should be $i(\varphi) \leftrightarrow i(\psi)$
$i(\varphi \wedge \psi)$ should be $i(\varphi) \wedge i(\psi)$
$i(\varphi \vee \psi)$ should be $i(\varphi) \vee i(\psi)$.

The case of the quantifiers '∀' and '∃' is a bit more complicated. At first glance, one might be tempted to say that

$i(\forall \alpha \varphi)$ should be $\forall \alpha i(\varphi)$
$i(\exists \alpha \varphi)$ should be $\exists \alpha i(\varphi)$.

Here we are just following the model of the preservation of logical form given above. Is this reasonable? We are requiring that if it is a theorem of Σ that every α satisfies φ, then it must be a theorem of Σ' that every α satisfies $i(\varphi)$. Would this prevent us from producing any interpretations which are desirable and legitimate?

Let Σ be a system of pure set theory — that is, a system which recognizes the existence only of sets. From the point of view of this system, sets are the only things in the universe. Now suppose Σ' is just like Σ except that it recognizes the existence of things (such as pencils and people) which are not sets. In Σ, a theorem of the form $\forall \alpha \varphi$ (where φ contains no quantifiers) will assert that everything in the universe satisfies the predicate φ. Corresponding to this in Σ' will be a theorem asserting that every *set* in the universe satisfies φ — that is, a theorem of the form $\forall \alpha (SET\alpha \rightarrow \varphi)$ where 'SET' is a unary predicate which means "———— is a set." Further, a theorem in Σ of the form $\exists \alpha \varphi$ (where, again, φ contains no quantifiers) will assert that something in the universe satisfies φ. Corresponding to this in Σ' will be a theorem asserting that some *set* satisfies φ — that is, a theorem of the form $\exists \alpha (SET\alpha \wedge \varphi)$. Given this information, it seems entirely reasonable to call a function i of the following sort an "interpretation of Σ within Σ'." Let i deal with the sentential connectives in the manner described above. But let i handle the quantifiers as follows:

$i(\forall\alpha\varphi)$ is $\forall\alpha(\text{SET}\alpha \to i(\varphi))$
$i(\exists\alpha\varphi)$ is $\exists\alpha(\text{SET}\alpha \wedge i(\varphi))$.

Since this function makes allowances for the wider universe of Σ' while leaving the theorems of Σ otherwise untouched, it supplies us with an interpretation of Σ within Σ' which is quite acceptable with regard to the preservation of meaning. And such an interpretation is illuminating because it shows how the results of pure set theory can be reproduced within applied set theory.

So the treatment of the quantifiers '∀' and '∃' first suggested above is too restrictive: it would unreasonably rule out desirable interpretations such as the one just suggested. Instead we say that

$i(\forall\alpha\varphi)$ should be $\forall\alpha(\psi \to i(\varphi))$
$i(\exists\alpha\psi)$ should be $\exists\alpha(\psi \wedge i(\varphi))$

where ψ is a formula such that $\exists\alpha\psi$ is a theorem of Σ' (so that we can infer $i(\exists\alpha\varphi)$ from $i(\forall\alpha\varphi)$). The addition of the formula ψ allows for interpretations like the one involving the predicate 'SET'.

So far, we have only discussed the effect of an interpretation function i on quantifiers and sentential connectives. In the case of predicates and variables, we require merely that i not cross between grammatical categories. That is, variables should be mapped into variables. And n-ary predicates (as discussed in chapter 7, §1) should be mapped into formulas with n argument places. To allow for the so-called contextual definition of class signs, definite descriptions, and such, we let i map expressions which function like proper names or operation symbols into expressions from other grammatical categories.

3. Abstraction as Interpretation

Predicates which are intended to express the relation of identity deserve particular attention. For it is philosophically important to distinguish between interpretations which map identity into identity and interpretations which map identity into a weaker relation. Usually the predicate '=' is used to stand for identity. So we want to distinguish between interpretation functions i such that

$i(\alpha=\beta)$ is $\alpha=\beta$

and interpretations i such that

$$i(\alpha=\beta) \quad \text{is} \quad \alpha \sim \beta$$

where '$\sim$' stands for some binary predicate other than '='. Interpretations of the latter sort are a form of definition by abstraction – an interpretation technique whose development has been guided most notably by Giuseppe Peano, Hermann Weyl, and Paul Lorenzen.[53] In the most general terms, abstraction is a restriction on one's language (or one's worldview) which allows a relatively weak relation to play the role of identity. Let's see how an interpretation function can perform this task.

Let i be an interpretation function which takes us from a system Σ into a system Σ' and which maps '=' into '$\sim$', where '$\sim$' denotes some relation weaker than identity. Σ will contain the theorem

$$\forall x \; x=x$$

(that is, everything is identical to itself). And, since i is an interpretation function, $i(\forall x \; x=x)$ will be a theorem of Σ'. Given the way we have decided to deal with the interpretation of quantifiers, $i(\forall x \; x=x)$ will have the form

$$\forall x (Fx \rightarrow x \sim x)$$

(that is, everything which has a certain property F stands in $\sim$ to itself). This means that if we consider only those things which have the property F, $\sim$ will appear to be reflexive. Furthermore, Σ will contain the theorem

$$\forall x \forall y (x=y \rightarrow y=x).$$

So Σ' will contain the theorem $i(\forall x \forall y(x=y \rightarrow y=x))$; that is,

$$\forall x(Fx \rightarrow \forall y(Fy \rightarrow (x \sim y \rightarrow y \sim x)));$$

which is equivalent to

$$\forall x \forall y ((Fx \land Fy) \rightarrow (x \sim y \rightarrow y \sim x)).$$

This means that if we consider only those things which have the property F, $\sim$ will appear to be symmetric. Furthermore, Σ will contain the theorem

$$\forall x \forall y \forall z ((x=y \land y=z) \rightarrow x=z).$$

So Σ' will contain the theorem

$$\forall x \forall y \forall z ((Fx \land Fy \land Fz) \rightarrow ((x \sim y \land y \sim z) \rightarrow x \sim z)).$$

This means that if we consider only those things which have the property F, $\sim$ will appear to be transitive. A reflexive, symmetric, and transitive relation is known as an "equivalence relation." So, to summarize what we have discovered so far, if, within Σ', we consider only things which have the property F, $\sim$ will appear to be an equivalence relation — that is, $\sim$ will appear to have several of the most important characteristics of identity. If, under certain conditions, $\sim$ also appears to satisfy Leibniz' Law, we would credit $\sim$ with a perfect impersonation of identity.

Every instance of Leibniz' Law (also known as the "principle of the indiscernibility of identicals") will be a theorem of Σ. That is, every formula of the form

$$\forall x \forall y ((Px \land x=y) \to Py)$$

will be a theorem of Σ. So every formula of the form

$$\forall x \forall y ((Fx \land Fy) \to ((i(Px) \land x \sim y) \to i(Py)))$$

will be a theorem of Σ'. This means that if we consider only things which have the property F and if we restrict ourselves to predicates of the form $i(Px)$ (that is, predicates which are the result of applying i to a predicate of Σ), then $\sim$ will appear to satisfy Leibniz' Law. So, under these conditions, $\sim$ will appear to have all the essential properties of identity. And, remember, definition by abstraction is essentially a matter of creating situations in which such an impersonation of the identity relation is possible. It is in this sense that interpretation can be a form of definition by abstraction.

To recapitulate: We perform abstraction when we restrict our language in such a way that a relatively weak relation is able to impersonate identity. The idea is that we pick out — or abstract — those forms of expression which do *not* allow us to see through the impersonation. The remaining forms of expression are "left behind" as residue.

This is a linguistically oriented reconstruction of the traditional notion that abstraction is a matter of selectively attending to certain features of an object while ignoring all the other features. For example, given some fixed modulus, the traditional view would have it that we obtain our idea of the congruence integer 7 by attending only to those features which 7 shares with all the integers congruent to it. As long as we maintain this state of selective attention (as long as we "abstract from" those features which enable us to discriminate between 7 and the integers to which it is congruent), 7 and its "congruence mates" will be indiscernible: they will appear to be fused into a single object which we christen "the congruence integer 7."

Thanks largely to some devastating criticisms offered by Gottlob Frege and Edmund Husserl, this psychologistic account of abstraction has fallen into disrepute.[54] Frege's alternative approach is to identify the congruence integer 7 with the set of all integers congruent to 7 (as we did in chapter 1). It is this which Max Black has in mind when he contends that abstraction presupposes a certain amount of set theory and, hence, cannot be used to explain what sets are. We see now, though, that there is an alternative to "abstraction as set formation." We can make sense of a modular arithmetic without treating congruence integers as sets of congruent integers. In fact, the ordinary old integers can themselves serve as congruence integers – as long as they are aided by an abstractive restriction which renders our language incapable of discriminating between congruent integers. Technically, this would involve our defining an interpretation function which maps the modular arithmetic into ordinary arithmetic while mapping identity into congruence. As outlined above, this would allow us to specify a sector of our ordinary arithmetical language within which congruent integers are indiscernible and, hence, within which congruence can successfully impersonate identity. The apparent fusion of congruent integers within the modular arithmetic could then be explained as an illusion created by our remaining within a restricted sector of our language. Since, within that sector, congruence has all the characteristics of identity, we treat it as identity. That is, we understand identity claims about congruence integers as follows: When we assert that $n=m$ within a modular arithmetic, we really mean that $n \equiv m$ (mod k) – and our interpretation function allows us to explain how we can get away with using the former (apparently stronger) locution in place of the latter. (By the way, since claims about the identity of congruence integers are to be regarded as disguised claims about the congruence of integers, the truth of the latter claims could certainly serve as a reliable criterion for the truth of the former. So "abstraction as interpretation" accords with the notion of abstraction which we touched on in chapters 1 and 2.)

Interpretation functions of the sort we have in mind here need not be regarded as set theoretic constructions (although, as a matter of fact, they sometimes are so regarded). For example, we could instead take them to be machine computable processes or rules for carrying out such processes. So Black is wrong to suppose that abstraction is irremediably set theoretic. (I think he is right, though, to insist that abstraction, however construed, will not supply us with a commonsense account of mathematical sets.[55])

4. Pseudo-Uniqueness

Assertions of uniqueness play an important role in mathematics. For example, to say that a relation R is a function is to say that, given any object x, there is a unique object y such that x stands in R to y (if, that is, x stands in R to anything at all). As in the case of other assertions of uniqueness, this can be regarded as a claim involving the identity relation:

$$\forall x \forall y \forall z ((Rxy \wedge Rxz) \rightarrow y=z).$$

Now let i be an interpretation function which maps '=' into '='. And let i' be an interpretation function which maps '=' into '~', where '~' denotes some relation weaker than identity. Then i will map the assertion that R is a function into the assertion that $i(R)$ is a function within a certain domain F. That is,

$$\forall x \forall y \forall z ((Fx \wedge Fy \wedge Fz) \rightarrow ((i(Rxy) \wedge i(Rxz)) \rightarrow y=z)).$$

But i' will give us

$$\forall x \forall y \forall z ((F'x \wedge F'y \wedge F'z) \rightarrow ((i'(Rxy) \wedge i'(Rxz)) \rightarrow y \sim z)).$$

That is, i' will map the assertion that R is a function into the assertion that $i'(R)$ is a pseudo-function within a certain domain F'. By a pseudo-function I mean a relation which appears to be a function merely because some other relation (in this case, ~) is successfully impersonating identity. In this sense, i' maps uniqueness claims into pseudo-uniqueness claims.

Whether such an interpretation is acceptable in a particular case will depend on one's philosophical views and on what one intends to *do* with the interpretation. A philosopher might use the existence of an interpretation as evidence that the interpreted theory can be dispensed with in favor of the theory in which it is interpreted. (This is what Daniel Bonevac means when he argues that certain interpretations can be regarded as ontological reductions.[56]) This move can be tempting when the interpreting theory is thought to be philosophically preferable in some way to the interpreted theory. Under such circumstances, many philosophers might not hesitate to abandon assertions about identity in favor of assertions about pseudo-identity. And those who do have qualms might find them outweighed by the desirability of dispensing with the interpreted theory. (For example, this theory, though useful in some ways, might assert the existence of entities which are repugnant to these philosophers.) On the other hand, a philosopher who regards mathematics as the study of certain abstract structures might balk at replacing claims about identity with claims about pseudo-identity. For this can entail that uniqueness becomes pseudo-uniqueness, functions become pseudo-functions, linear orderings become pseudo–linear-orderings, continua become pseudo-continua, and, in general, theorems about structures of a particular sort become theorems about pseudo-structures-of-that-sort. A mathematical structuralist might regard this as a subversion of mathematics. Be that as it may, I wish merely to point out that there is a philosophically significant difference between interpretations which map identity into identity and interpretations which map identity into pseudo-identity. That is, there may be cases in which the appropriateness of interpretations of the former sort is not questioned, but the appropriateness of interpretations of the latter sort is open to philosophical debate.

This concludes our first logical interlude. We are now in a position to examine an important philosophical account of set theory which will be more intelligible to us now that we have some grasp of formal interpretability.

V

FORMALISM

> In the Realist philosophy, one wholeheartedly accepts traditional mathematics at face value. All questions such as the Continuum Hypothesis are either true or false in the real world despite their independence from various axiom systems. The Realist position is probably the one which most mathematicians would prefer to take. It is not until [the average mathematician] becomes aware of some of the difficulties in set theory that he would even begin to question it. If these difficulties particularly upset him, he will rush to the shelter of Formalism.... [M]athematics may be likened to a Promethean labor, full of life, energy and great wonder, yet containing the seed of an overwhelming self-doubt. — Paul J. Cohen[57]

1. Universal Theories as Mathematical Objects

Crudely put, formalism as a philosophy of mathematics is the view that the proper and, indeed, only possible objects of mathematical research are mathematical *theories* (rather than the items to which those theories purportedly refer). In response to a host of disconcerting interpretability results, formalism has become a popular position among set theorists. It is an indisputable fact that set theorists do largely concern themselves with theorems about set *theories* (as opposed to theorems about *sets*). So a formalist outlook would be consistent with the everyday practice of many set theorists. It hardly follows, though, that such an outlook is consistent with the best available philosophical and logical insights into mathematics. The logical and philosophical viability of formalism as a philosophy of set theory shall be our concern in this chapter.

Let us begin by sketching the basic issues from the point of view of a practicing set theorist. The specialized mathematical discipline known as "set theory" is, in one of its guises, the study of those formal systems which are *universal* in the sense that all mainstream mathematical systems can be interpreted within them and which, furthermore, are *set theoretic* by virtue of (apparently) describing a binary relation $\in$ which satisfies the axioms of Zermelo Set Theory. The notion that the primary objects of set theoretic study are not sets but set *theories* is a common source of comfort for mathematicians when they are pestered by philosophers and philosophically minded colleagues. (Unwelcome questions from which a mathematician might seek safe haven include: "What are sets?" "How can you know anything about mathematical sets given that they must lie outside of time and space?"[58] "How can you justify your use of nonconstructive proof techniques?" And, as we shall see, "What position does $2^{\aleph_0}$ occupy in the sequence of infinite cardinals?") The appeal of formalism could only be enhanced by the recent history of set theory. The major developments in twentieth-century set theory are most naturally described as discoveries about the properties of and relations between set theoretic *deductive systems*. For example, the towering work of Kurt Gödel and Paul Cohen established that a formula known as the "Continuum Hypothesis" is neither provable nor refutable in any set theoretic deductive system which is both consistent and currently uncontroversial. This is a striking advance in our knowledge of set theories. But, at the same time, it highlights our ineptitude at answering important questions about the alleged *objects* of those theories. To be told that our available mathematical resources are insufficient to prove or refute an extremely fundamental set theoretic claim is (it would seem) to learn little about *sets,* but quite a bit about our ignorance of which sets exist and what structures they form.[59] How natural it is, then, to concentrate on the items with which we can feel smugly intimate — namely set theories regarded as themselves important objects of mathematical study. This position has more than convenience to recommend it: its philosophical respectability has been upheld by as weighty a figure as Edmund Husserl.[60]

Husserl (1859–1938), a mathematician converted to philosophy, emphasized both the meta-theoretic and the universal character of set theory. The notion of an absolutely *universal* theory is a regulative idea governing much of the research into the foundations of mathematics. But some thinkers have little patience for the distant charms of a Kantian *telos*. Not content merely to pursue an unrealizable ideal, mathematicians frequently claim that some particular formulation of set theory is universal. For example, Kenneth Kunen declares that, "in axiomatic set theory we formulate a few simple axioms . . . from [which] all known mathematics may be derived."[61] Construed uncharitably, this claim is simply false. Taken in the right spirit, however, its truth will be generally acknowledged. While the concepts of "theory" and "interpretability" at work here are rigorously formal, the appropriate concept of "universality" is utterly informal and irredeemably vague (but not unacceptable or unintelligible). When one declares that a particular theory is an adequate basis for the interpretation of "all known mathematics" or "all standard mathematical systems" or "all mainstream mathematical theories," one assumes that an informed consensus can be reached about which formal systems are more or less central to mathematics as we currently know it and which are more or less peripheral. The "universe" within which a theory is claimed to be universal will, it is assumed, not extend far into the periphery of contemporary mathematics. A set theory which could not account for the results of classical analysis would have no claim to universality. But, one which was too weak to reproduce the deductive structure of an eccentric logician's pet theory could nonetheless be counted as universal. If, in the course of time, the eccentric's theory won a place for itself in the mathematical mainstream, the criteria for universality would alter accordingly. In their pursuit of an absolutely universal theory, set theorists guarantee by their own activities that even the "relative universality" of their creations will fade. Systems which successfully encompass the core components of mathematics tend to become central themselves and, by thus expanding and enriching the region calling for circumscription, bring forth ever more powerful theories at the periphery.

2. Cantor's Continuum Hypothesis

We have indicated that set theory (the discipline) is viewed by many of its devotees as an investigation into the properties not of sets, but of ("relatively") universal set theories. We must now examine more closely the phenomenon which has most powerfully drawn set theorists toward this formalist outlook. I refer to the independence of the Continuum Hypothesis from the currently accepted set theories and, even worse, from all known set theories which have hitherto seemed likely to become accepted.

We saw in chapter 1 that Georg Cantor constructed a transfinite arithmetic. This involved his showing that the infinite cardinal numbers form a sequence of the same basic type as the sequence of natural numbers. (Both sequences are well-orderings—a notion to be defined in chapter 8.) He also showed how arithmetical operations such as exponentiation can be extended into the transfinite. Thus, Cantor supplied us with two ways of characterizing an infinite cardinal: as an occupant of a particular position in the sequence of cardinal numbers and as the result of a particular application of an arithmetical operation. It turns out that the number of points on the real line (also known as the "linear continuum") can easily be characterized in the latter way: they are $2^{\aleph_0}$ in number ($\aleph_0$ or "aleph null" being the first infinite cardinal). One would, of course, also like to be able to characterize this number in terms of its position in the sequence of infinite cardinals. One would like to be able to announce that the number of points on the linear continuum is, say, $\aleph_2$ (the third infinite cardinal). If infinite cardinals were as well behaved as finite ones, we would simply *calculate* that $2^{\aleph_0}=\aleph_\alpha$ for some ordinal α (just as we would calculate that, say, $2^7=128$). We would then know that the points on the linear continuum are $\aleph_\alpha$ in number. Sadly, the infinite cardinals are (given our current resources) extraordinarily ill behaved. We simply are not in a position to make such a calculation.

In the absence of an appropriate technique of calculation, Cantor tried his hand at *guessing*. According to his celebrated Continuum Hypothesis (CH) of 1878, the points on the linear continuum are $\aleph_1$ in number or, equivalently, $2^{\aleph_0}=\aleph_1$. This would mean that infinite sets of points come in exactly two sizes: $\aleph_0$ (the cardinality of the natural numbers) and $\aleph_1$ (the alleged cardinality of the whole real line). According to the Generalized Continuum Hypothesis (GCH), $2^{\aleph_\alpha}=\aleph_{\alpha+1}$ for any ordinal α. In the beth notation of chapter 1, GCH states that $\beth_\alpha=\aleph_\alpha$.

In 1938, Kurt Gödel (1906–1978) established that GCH and *a fortiori* CH cannot be refuted in the standard axiomatizations of set theory unless those axiomatizations are inconsistent.[62] More particularly, he proceeded as follows. Gödel discovered a formula known as the "Axiom of Constructibility" (abbreviated as 'V=L') which, when appended to the standard set theories, allows one to prove GCH. For example, GCH is provable in GB+V=L (Gödel-Bernays Set Theory augmented by the Axiom of Constructibility). Gödel also showed that GB+V=L is interpretable in GB. As we saw in chapter 4, this means that the consistency of GB+V=L follows from the consistency of GB. Now suppose GB is consistent. Then GCH cannot be disproved in GB. For, otherwise, GCH could be both proved and disproved in GB+V=L and, hence, GB+V=L would be inconsistent. But GB+V=L must be consistent if GB is. Thus the consistency of GB implies the irrefutability of GCH in GB.

The Axiom of Constructibility states that all the sets in the universe have a certain property (known as – you guessed it! – "constructibility") which allows us to place them in a well-ordered array. This is sometimes expressed by saying that V, the class of all mathematical sets, is identical to L, a class consisting of certain peculiarly well-behaved sets (hence the expression 'V=L'). If it were mathematically plausible that V=L, our troubles with CH and GCH would be at an end. Gödel's proof of GCH within GB+V=L would establish the truth of both GCH and CH. Unfortunately, the Axiom of Constructibility has won the hearts of few, if any, set theorists. Kenneth Kunen's observation seems typical: ". . . . there is no reason to believe that all mathematical objects . . . lie in L. Thus, we do not consider V=L to be a plausible basic axiom . . . we merely use it as a tool to obtain relative consistency results."[63]

In 1963, Paul Cohen established that CH and *a fortiori* GCH cannot be proved in the standard axiomatizations of set theory unless those axiomatizations are inconsistent.[64] More particularly, via his ingenious and extremely fruitful technique of "forcing," he constructed a function which would map any contradiction in ZFC+ ¬CH into a contradiction in ZFC (ZFC+ ¬CH being Zermelo-Fraenkel Set Theory plus the Axiom of Choice and the *negation* of the Continuum Hypothesis). It follows that ZFC+ ¬CH is consistent if ZFC is. And this, in turn, implies that CH is not provable in ZF unless ZF is inconsistent (ZF being Zermelo-Fraenkel Set Theory *without* the Axiom of Choice).[65] We are able to make the jump from ZFC to ZF because (thanks to Gödel) we know that the latter is consistent only if the former is. (For suppose that CH is provable in ZF. Then, *a fortiori*, CH is provable in ZFC. So ZFC+ ¬CH is inconsistent. It follows, by Cohen's result, that ZFC is inconsistent. So, by Gödel's result, ZF is inconsistent. As desired, we have shown that CH is provable in ZF only if ZF is inconsistent.)

Gödel and Cohen's results combined tell us that CH and GCH are undecidable (i.e., neither provable nor refutable) in the currently accepted set theories. (The Gödel result for GB can easily be transferred to ZF and other popular theories. Similarly, the Cohen result can easily be transferred to GB *et alii*.) Thus, the position of $2^{\aleph_0}$ in the sequence of infinite cardinals is undetermined by current assumptions about sets. Cohen has, furthermore, shown that it is *extravagantly* undetermined. As far as the theories now in vogue are concerned, $2^{\aleph_0}$ could be $\aleph_n$ or even $\aleph_{\omega+n}$ for *any* finite ordinal n greater than 0. In fact, even stronger indeterminacy results are available.[66] What we, as philosophers, must now determine is whether these technical discoveries reflect unfavorably on our grasp of the concept "mathematical set."

3. The Robinson-Cohen Challenge

According to the logician Abraham Robinson, the independence of CH and GCH suggests "... that the entire notion of the universe of sets is meaningless"; and he concludes that "... the present situation in Set Theory favors the Formalist."[67] Paul Cohen was so impressed by Robinson's advocacy of formalism, that he, too, leapt on the band wagon (though he admitted discomfort with "... the admission that CH, perhaps the first significant question about uncountable sets which can be asked, has no intrinsic meaning").[68] I have been careful to speak rather vaguely of Robinson's "advocacy" of formalism, for it is unclear what *argument*, if any, he and Cohen mean to be giving. This has left their position open to vicious caricatures.[69] Our own philosophical (as opposed to purely historical) interests will best be served if we ascribe to Robinson and Cohen the strongest defense of their position we can muster – even if this involves some distortion of their actual intentions.

Some of Robinson's contentions are clear enough:

1. Our concept of mathematical set is hopelessly defective.
2. Set theoretic formulas are, in general, meaningless.
3. We should treat such formulas as voiceless mathematical objects rather than as articulate propositions *about* mathematical objects.

Theses 2 and 3 are meant to be supported by Theses 1 and 2 respectively, while Thesis 1 is meant to be supported by the current undecidability of CH and GCH. Let us concentrate on this last-mentioned step.

The bare independence of CH and GCH from the standard set theories does not, in itself, imply Thesis 1 – or even render it particularly plausible. Nonformalists have acknowledged for decades that the usual axiomatizations fail to capture adequately their concept of mathematical set. (This is clear, for example, from Gödel's proof that the axioms of GB all come out true in the constructible universe L.) So, many nonformalists have looked forward complacently to an endless series of more and more adequate characterizations of their set concept. (This is just one aspect of the inevitable "relativity" of set theory's universality.) From this point of view, independence results are only to be expected and are to be regarded merely as invitations to strengthen our current theories. Indeed, an independence proof might even suggest to us how better to articulate our concept of set and, thus, would be most welcome.

If, as Cohen maintains, "The recent independence results are [a] challenge to the Realist position,"[70] this must be because either the proofs or the propositions whose independence is proved have special features. And, to a certain extent, they do. First of all, the proofs of Gödel and Cohen seem not to have guided us even vaguely in the direction of an appropriate strengthening of our theories. Secondly, our experience with ordinary arithmetic makes the undecidability of CH seem particularly disastrous. (Mustn't an arithmetic be fundamentally ill conceived if, within it, an elementary application of an elementary operation yields a wildly undetermined result?[71]) Finally, in the two and a half decades since Cohen's breakthrough, concerted efforts at determining the truth or falsity of CH and GCH have not been crowned with success (though the continuum problem has inspired much research which is valuable in other ways). To a pessimist, this might suggest that such efforts are futile. And a natural explanation for this futility would be that CH, GCH, and their ilk are neither true nor false (because they are not meaningful propositions).

We have just sketched what is, at best, a probabilistic argument for formalism. For all we know, within the next few years a new axiom which decides CH and GCH will be discovered and a consensus will develop that this new axiom accords with a widely held concept of mathematical set.[72] Since this would utterly obliterate the Robinson-Cohen route to formalism, the plausibility of their argument must depend in part on our belief that such a development is unlikely. But, at the moment, we could be led to this belief by nothing more dependable than an informed assessment of probabilities. On the other hand, until nonformalists are able to articulate their concept of set in a way which renders CH and GCH decidable, the Robinson-Cohen argument can be expected to haunt them as "the seed of an overwhelming self-doubt" (to use Cohen's phrase).

Although Robinson and Cohen have not offered a compelling argument for formalism, they have nonetheless performed a valuable service. By raising the issue in a provocative (one might even say irritating) way, they have made the *absence* of a convincing treatment of formalism intolerable. To their nonformalist colleagues they have issued the challenge: show us that you possess a determinate concept of mathematical set by articulating it in a way which settles fundamental questions (such as the status of CH and GCH). To philosophers they have, in effect, issued the challenge: lessen the tension produced by our lack of such an articulation by establishing the legitimacy or illegitimacy of formalism in a way which does not depend upon a questionable assessment of probabilities. We shall now

72 FORMALISM

turn to a philosophical argument of roughly the sort demanded. First, however, let me note that we may *already* possess set theories which decide CH and GCH and which *ought* to be widely hailed as faithful expressions of our standard conception of mathematical sets. We shall be in a position to pursue this claim only at the end of chapter 7.

4. A Dummettian Argument

The following argument owes its essential features to the philosopher Michael Dummett.[73] I should note, however, that Dummett is not in the habit of exercising his philosophical gifts in the service of *formalism*. So our Dummettian argument will not have an entirely Dummettian conclusion.

Definition: Our knowledge of the meaning of an expression φ is *implicit* if and only if that knowledge does not consist entirely in a capacity to state φ's meaning.

Example: Every competent speaker of English knows the meaning of the definite article 'the'. But few (if any) such speakers are able to *state* the meaning of 'the'. (Just try! On one plausible view of meaning, this would involve stating the contribution which 'the' makes to the meaning of each sentence in which it occurs.) An ordinary speaker's knowledge of the meaning of 'the' consists in knowing *how* to use 'the' correctly rather than in knowing *that* 'the' means such-and-such. This knowledge is said to be implicit.

Claim 1: If we know the meaning of a mathematical expression implicitly, then that knowledge consists in our possession of mental capacities which can be publicly exhibited.

Justification: At stake here is the communicability of mathematics. As Dummett writes: "To suppose that there is an ingredient of meaning which [cannot be publicly exhibited] is to suppose that someone might have learned all that is directly taught when the language of mathematical theory is taught to him, and might then behave in every way like someone who understood the language, and yet not actually understand it, or understand it only incorrectly. . . . A notion of meaning so private to the individual is one that has become completely irrelevant to mathematics as it is actually practised, namely as a body of theory on which many individuals are corporately engaged, an enquiry within which each can communicate his results to others."[74]

Claim 2: If we know the meaning of a mathematical expression implicitly, then that knowledge consists in our grasp of the distinction between correct and incorrect uses of that expression.

Justification: By Claim 1, our implicit knowledge of what a mathematical expression means must be publicly exhibitable. How does one go about doing this exhibiting? That's easy: by using the expression correctly. It does not immediately follow that our knowledge of meaning *consists* in our grasp of proper usage. But since that knowledge is adequately exhibited through proper use, it's hard to see what else the knowledge could amount to.

Claim 3: If we know implicitly what a mathematical expression φ means, then φ's meaning is that by virtue of which an employment of φ is correct or incorrect.

Justification: If our knowledge of the meaning of x consists in our grasp of y, then y is the meaning of x. (More briefly: If to know the meaning of x is to know y, then y is the meaning of x.) So Claim 3 follows from Claim 2.

Postulate: The meaning of a declarative sentence determines the conditions under which that sentence is true.

Discussion: I don't mean to suggest that the truth conditions of a declarative sentence will always be terribly clear. The point is simply that to the extent those conditions are determined, they are determined by the *meaning* of the sentence. I have called this a "postulate" because it strikes me as too basic to admit of justification. (Or am I just too unimaginative to dream up a justification?)

Claim 4: If we know implicitly what a mathematical sentence φ means, then φ's truth conditions can be no better defined than the distinction between correct and incorrect uses of φ.

Justification: If we assume that truth conditions can be no better defined than the meanings which fix them, then Claim 4 follows from our Postulate and Claim 3.

Claim 5: If φ is a set theoretic sentence containing only '∈' and logical vocabulary, then φ's truth conditions can be no better defined than the distinction between correct and incorrect uses of φ.

Justification: The view of mathematics sketched in chapter 1 suggests that it is preeminently *set theoretic* expressions whose meanings we know implicitly, for mathematical expressions which are not overtly set theoretic are to be explicated via set theoretic interpretations. To state the point somewhat differently, since all of mainstream mathematics is interpretable in the standard set theories, our search for implicitly known meanings can take place entirely within the language of set theory. This will simplify our task immensely: a set theory requires for its non-logical vocabulary no more than the binary relation symbol '∈' (epsilon). So the crucial question is whether we know what '∈' means only implicitly.

We are accustomed to proclaim glibly that '∈' stands for the relation of "set membership." Before embarking on our investigations in chapters 2 and 3, we might even have claimed that our knowledge of epsilon's meaning consists in our ability to *state* its meaning in just this way. We now realize that '∈' cannot stand for any relation which we are in a position to understand prior to mastering mathematical set theories. To tell beginners baldly that '$x \in y$' means "x is a member of the set y" probably has some heuristic value – but we, as philosophers, must remind ourselves that we can grasp the notions of set and membership at work here only if we acquire technical, set theoretic competence. Unless it can be shown that this competence acquaints us with the meaning of epsilon precisely by equipping us to *state* that meaning (which seems not in fact to be the case), we must suppose that our knowledge of epsilon's meaning is implicit.

If our knowledge of epsilon's meaning is implicit, what may we conclude about our knowledge of what set theoretic *sentences* mean? For example, consider the sentence '$\neg \exists x \ x \in x$' (that is, informally, "nothing is a member of itself"). Certainly, a *necessary condition* for knowing the meaning of this sentence is knowing the meaning of '∈'. But we must determine whether our knowledge of what this sentence means is partly *constituted* by our knowledge of what '∈' means. This would entail that our knowledge of what '$\neg \exists x \ x \in x$' means consists partly in our ability to use '∈' correctly – so this knowledge would be implicit. But here we come upon an odd realization: We, as set theorists, employ '∈' only within sentences – hence, to use '∈' correctly is to deal properly with sentences containing '∈'. It follows that our knowledge of what '∈' means consists partly in our ability to use '$\neg \exists x \ x \in x$' correctly. But then how could our knowledge of what the latter means consist partly in our ability to use the former correctly? Doesn't this lead to a vicious circle?

Not really. The moral is actually that our knowledge of what '∈' means and our knowledge of what set theoretic sentences mean are interdependent. We do not first acquire an abstract knowledge of what '∈' means and only then figure out what sentences containing it mean. Rather, we acquire our implicit knowledge of epsilon's meaning progressively by gaining greater and greater mastery over set theoretic sentences. As our set theoretic competence grows, so does our right to claim a grasp of '∈'. And as our grasp of '∈' is strengthened, so is our right to claim a grasp of set theoretic sentences. By correctly employing a host of set theoretic sentences within a variety of proofs and definitions, we display an appreciation for the contribution which '∈' makes to the meaning of any given set theoretic sentence (say, '$\neg \exists x \ x \in x$'). And this, coupled with our mastery of the logical vocabulary ('∃', '¬', '→', etc.), exhausts our knowledge of what set theoretic sentences mean—or, at least, it does so in the case of those sentences containing only '∈' and logical vocabulary. There is no further requirement that we be able to *state* what those sentences mean; nor is it clear that anyone could produce such a statement. We conclude that our knowledge of what the basic set theoretic sentences mean is implicit. Hence Claim 5 follows from Claim 4. So if the distinction between correct and incorrect uses of set theoretic sentences is ill defined, then the truth conditions of those sentences are correspondingly ill defined.

Claim 6: If the distinction between adequate and inadequate set theoretic proofs is unclear, there will be a corresponding unclarity in the distinction between correct and incorrect uses of set theoretic expressions.

Justification: Proofs are the primary arena for the employment of set theoretic expressions. (Set theorists also construct definitions and perform calculations; but since correct mathematical definitions are in principle eliminable and since calculations can be recast as proofs, we may concentrate entirely on proofs.) To use a set theoretic expression correctly is, primarily, to use that expression in a way which can contribute to the adequacy of a set theoretic proof. To use a set theoretic expression incorrectly is, primarily, to use that expression in a way which can render a set theoretic proof inadequate. If we are unclear about the distinction between adequate and inadequate set theoretic proofs, we shall be correspondingly unclear about the contributions which particular employments of set theoretic expressions make to the adequacy or inadequacy of proofs. So if we are unclear about the distinction between adequate and inadequate set theoretic proofs, we shall be correspondingly unclear about the distinction between correct and incorrect uses of set theoretic expressions.

Claim 7: If φ is a set theoretic sentence containing only '∈' and logical vocabulary, then φ's truth conditions can be no better defined than the distinction between adequate and inadequate proofs of φ.

Justification: Claims 5 and 6 imply that unclarities in the distinction between adequate and inadequate set theoretic proofs will be reflected in deficiencies in the determination of set theoretic truth conditions. What form might these deficiencies take? The crucial point to note here is that an adequate mathematical proof supplies a tacit delineation of truth conditions for its conclusion. If we suppose that set theoretic sentences are capable of being true or false (if we suppose that set theoretic sentences are *not* meaningless), then an adequate set theoretic proof will be one that establishes the truth of its conclusion. In the course of doing this establishing, the proof will mark out features of the set theoretic universe by virtue of which the conclusion is true. Conversely, unclarity about what should count as a proof of a set theoretic sentence would signal a corresponding unclarity about the conditions which would render the sentence true. We conclude that if φ is a set theoretic sentence containing only '$\in$' and logical vocabulary, then set theoretic proof practices fix φ's truth conditions by fixing what is to count as a proof of φ. If the concept "proof of φ" is indeterminate, then φ's truth conditions will be at least as indeterminate.

Claim 8: CH is neither true nor false.

Justification: Set theories such as ZFC and GB are canonical representations of actual set theoretic proof practices. The formal undecidability of CH within these theories is regarded as decisive evidence that the informal proof techniques currently in use are unable to supply us with an informal proof or disproof of CH. It follows that overt mathematical practices would have to be augmented in order for them to draw a clear line between adequate and inadequate proofs of CH. Current practices place limits on admissible augmentations. Still, they leave what is to count as a proof of CH significantly undetermined. By Claim 7, the truth conditions of CH are at least as undetermined. This does not mean that we have failed to *discern* the truth value of CH. It means rather that *what it is* for CH to be true or false is significantly undetermined. A sentence is true by virtue of certain specified conditions being satisfied. Since, in the case of CH, there is no more than a partial specification of such conditions, CH is currently not the sort of sentence which is capable of being true. Similarly, CH is currently not the sort of sentence which is capable of being false. So CH is neither true nor false. It does not (immediately) follow that CH is *meaningless,* but merely that if CH is endowed with meaning, it is not endowed in such a way as to render it a fit bearer of truth or falsity.

Claim 9: If set theoretic sentences are to be regarded as meaningful, the logical vocabulary of classical set theories must be interpreted in a nonstandard way.

Justification: We have in mind here set theories such as ZFC and GB whose underlying logic is classical. A distinctive component of classical logic is the law of excluded middle (LEM): the principle that $(\varphi \vee \neg \varphi)$ is a logical truth no matter what sentence φ is. On the standard reading of '$\vee$' and '$\neg$', a disjunction is true just in case one of its disjuncts is true and negations are true just in case the sentence being negated is false. On this reading, $\neg$CH is not true (since, by

Claim 8, CH is not false) and, thus, (CH ∨ ¬CH) is not true either (since, again applying Claim 8, neither of its disjuncts are true). So *a fortiori* (CH ∨ ¬CH) is not a *logical* truth and, in this setting, LEM is not a reliable logical principle. Hence, given a standard reading of the logical constants, it is not appropriate to employ classical logic within set theory. If, therefore, we are to retain classical set theories such as ZFC and GB, we either must interpret their logical vocabulary in a nonstandard way or must abandon the task of interpretation altogether.

Claim 10: If no defensible nonstandard interpretation of set theory's logical vocabulary can be found, there will be no viable alternative to supposing that classical set theoretic sentences are meaningless.

Justification: This follows directly from Claim 9.

As the conditional nature of Claim 10 indicates, we have not decisively established the philosophical inevitability of formalism. Instead, we have supplied ourselves with a program for rendering formalism progressively more inviting (at least among those with a commitment to classical set theory). Although a formalist can hardly enumerate and then discredit all *possible* nonstandard approaches to the semantics of classical set theory, the formalist can deal with such approaches as they arise; and an uninterrupted string of successful formalist assaults could leave the antiformalist camp with little credibility intact. We ourselves shall examine and criticize just one antiformalist approach to the semantics of set theory. This will serve merely as an example of a single contribution to an ongoing program.

Why exactly is it inappropriate to give the logic of classical set theory a standard interpretation? One antiformalist view is that the *richness* of our conception of set is what undermines the standard semantics. The idea is that we possess an abundance of pictures of the set theoretic universe no one of which can be singled out as the honest-to-goodness true picture. From this point of view it is foolish to make truth *simpliciter* the central notion of our semantics – for we are not in a position to say what *the* truth about sets is. We must instead employ the relativized notion of truth-in-a-picture. For example, whereas we lack adequate grounds for asserting that CH is *absolutely* true or false, we can make well-grounded claims about whether CH is true or false in a particular picture of the set theoretic universe. Set theoretic theorems are those sentences which are true in *every* picture; non-theorems (such as CH) are those sentences which are true in some pictures and false in others. A disjunction is to be regarded as true-in-a-picture as long as one of its disjuncts is true in that picture. A negation is to be regarded as true-in-a-picture as long as the sentence being negated is false in that picture. Each picture is to be regarded as entirely definite (our problem is supposed to be that we have *too many* pictures, not that any of those pictures have cloudy contents); so each set theoretic sentence is to be regarded as either true or false in each picture. Hence, given this nonstandard approach to semantics, (CH ∨ ¬CH) is *rightly* counted as a theorem, because in every picture either CH is true or ¬CH is true and, hence, in every picture (CH ∨ ¬CH) as a whole is true.[75]

Here is a possible formalist response. Suppose (for the sake of argument) that we are able to grasp a fully definite picture of a set theoretic universe in which, say, CH is true. Call this picture 'Π'. How is our grasp of Π to be publicly exhibited? (Remember, we have already decided that mathematics is a communal activity within which irredeemably private concepts have no place.) Since proofs are the primary arena for the employment of set theoretic expressions, we can expect our grasp of Π to consist in our mastery of publicly exhibitable proof practices – in particular, practices which allow us to discriminate between sentences which are true-in-Π and sentences which are false-in-Π. If such proof practices actually exist, it should be possible to formulate a theory which adequately captures their essential features (just as ZFC and GB adequately capture the essential features of the proof practices which pick out sentences which are true-in-every-picture). If Π were a picture of Gödel's constructible universe L, then GB+V=L would be the natural candidate for our canonical theory. If Π were a picture of some other universe in which CH holds, then we would be tempted to adopt some other augmentation of GB or ZFC. In either case, we would run up against the same problem: Gödel has shown that *every* augmentation of GB or ZFC will leave some sentence or other undecidable (as long as the axioms of the augmented theory and any applications of its rules of inference can be mechanically recognized). For example, even though CH is provable in GB+V=L, it is possible to identify another sentence φ which is neither provable nore refutable in GB+V=L (just as CH is neither provable nor refutable in GB). By an argument strictly analogous to our justification of Claim 8, we can establish that such a φ is neither true-in-Π nor false-in-Π (since deficiencies in our canonical theory are signs of identifiable deficiencies in our grasp of Π and, hence, are signs of identifiable indeterminacies in the notion of truth-in-Π). So we were wrong to suppose that our grasp of Π is fully determinate. More generally, it is fanciful to suppose that our conception of set is overly rich. We do not possess an overabundance of fully definite pictures of the set theoretic universe; in fact, we possess no such pictures at all.

5. Why Bother?

According to the view we are entertaining in this chapter, the objects of set theoretic research are universal set theories. These theories are to be regarded solely as mathematical structures – not as significant characterizations of mathematical structures. Yet our utterances *about* universal set theories *are* meant to be significant. When we say that GB+V=L is interpretable in GB, we take ourselves to be asserting something both meaningful and true. We must take care, then, that our metamathematics (our theory of mathematical theories) is immune to the difficulties of the previous section. We must take care lest unclarities in the distinction between correct and incorrect uses of metamathematical expressions force us to regard metamathematical assertions as meaningless.

Even if we managed to avoid this pitfall (and there is good reason to think it *can* be avoided), challenging questions would remain: Why bother? If Cantorian set theories are meaningless, why keep fiddling with them? Why not just trash them?

An obvious first response is to maintain that these questions are inappropriate and even bizarre. Formalists view set theories as complex mathematical objects, not as collections of propositions. A sentence whose meaninglessness has been revealed should certainly not be treated as a significant assertion (if I may be forgiven a virtually tautological observation) – but this in itself does not establish that the sentence is unfit to serve as an object of mathematical study. (Should an algebraist be deterred by the revelation that finite simple groups are "meaningless"?) Whether a line of mathematical research is worth pursuing is something for *mathematicians* to decide. We armchair observers should try to keep our pontifical pronouncements on the matter to a minimum.[76]

Still, we would be most interested to know why someone better qualified than our humble selves might find the formalistic study of set theories to be of interest. Allow me to cite a few reasons.

1. Thanks to a device known as "Gödel numbering," we can show that metamathematical assertions about set theories are equivalent to propositions of *arithmetic*. So, for example, the assertion that CH is unprovable in ZF if ZF is consistent is equivalent to a family of

assertions about natural numbers. By proving the former assertion, we also establish the truth of the latter. It follows that if all arithmetical knowledge is worth pursuing, then so is metamathematical knowledge about set theories.[77]

2. The great mathematician and arch-formalist David Hilbert (1862-1943) hoped to show that certain powerful and tractable theories which he considered meaningless are conservative extensions of the relatively weak and unwieldy theories which he considered meaningful. Let's see what this means. Let T be, say, a weak formulation of arithmetic in the language L. Let L' be the result of adding new symbols to L. And let T' be the result of adding to T new axioms which are formulated in the expanded language L'. The theorems of T' will then be of two sorts: those which contain symbols peculiar to the new language L' and those which can be formulated using just the resources of L. If the theorems of the latter sort are precisely the theorems of T, then T' is a conservative extension of T. The point is that T' does not supply us with any theorems expressible in L which were not already granted us by T. This means that T' could be used as a dependable tool for proving theorems of T (in this case, theorems of arithmetic) regardless of what sort of theory T' is. For example, T' could be an infinitary set theory which formalists regard as meaningless. Set theory might, under these circumstances, provide us with reliable short cuts to number theoretic truths; we might be able to substitute elegant set theoretic proofs for tortuous arithmetical ones. (Something very much like this actually does occur in analytic number theory.) This sort of conservative extension result would be another way of basing the mathematical importance of set theory on that of arithmetic. Although, for reasons I won't detail, Gödel's Second Incompleteness Theorem makes it seem unlikely that any sweeping result of this sort is obtainable,[78] we might yet hope for some more modest advances. Metamathematical investigations have already yielded conservative extension results for narrowly defined classes of formulae.[79] And appropriate advances of this sort might justify our continued interest in the properties of set theories. For we might be able to show that Cantor's offspring guide us elegantly and reliably to important truths.

3. As we saw in chapter 1, set theory has inspired the crossfertilization of mathematical disciplines which might otherwise have maintained a barren self-involvement. If the meaningfulness of set theoretic sentences cannot ultimately be affirmed, this will not erase set theory's past success as a mathematical adhesive – nor will it tarnish its future prospects. If Cantorian set theories are shown to be

meaningless, it will thereby be established that meaningless theories can have immense heuristic value. Isn't this reason enough to investigate the properties of ZF, GB, and their kin?

* * *

I have tried to show that set theoretic formalism is a genuinely compelling philosophical outlook. In our next chapter, we shall learn more concretely what is involved in treating set theories as mathematical objects. In the remaining chapters, we shall consider whether formalism has any serious philosophical competition.

VI

SOME METAMATHEMATICS

1. A Formal Language and Its Logic

Chapter 4 featured a rather abstract discussion of interpretation functions and chapter 5 was dotted with vague references to "formal deductive systems." We shall now descend to a more concrete level. Having familiarized ourselves with a particular set theoretic deductive system, we shall work through a detailed interpretability proof.

The first step in describing a formal theory is a definition of the language within which that theory is expressed. The logical vocabulary of our language will consist of the universal quantifier '$\forall$', the identity symbol '=', the sentential connectives '$\rightarrow$' and '$\neg$', the parentheses ')' and '(', and a collection of variables ('x', 'y', 'z' among others). The non-logical vocabulary will, first of all, feature just one relation symbol: namely, '$\in$'. A relation symbol is, naturally, a symbol which expresses a relation, and '$\in$' is normally meant to express the relation of set membership. That is, a formula of the form '$x \in y$' is supposed to express that x is a member of the set y. Of course, those of us who have been converted to formalism by the arguments of chapter 5 will deny that '$\in$' expresses any relation at all. And even the non-converts will have been convinced by chapters 2 and 3 that casual appeals to commonsense set talk can be hazardous in a mathematical setting. I offer the above "explanation" of '$\in$' only because it has been found to have some heuristic value.

Our language will also contain set operators of the form $\{\alpha: \varphi\}$ — where α is a variable and φ is a formula of our language. A statement of the form $z \in \{x: \varphi\}$ is supposed to express that z is a member of the set consisting of all the objects which satisfy the formula φ. For example, '$z \in \{x: x \neq y\}$' is supposed to say that z is a member of the set consisting of all the objects which are not identical to y. (Again, it would be a mistake to think that this "explanation" genuinely reveals what set operators mean. Nonetheless, it is thought to be valuable for initiates to hear incantations of this sort.) For technical reasons, it is desirable that our language also contain a collection of individual constants ('a', 'b', 'c' among others) which will function grammatically like proper names.

At the risk of seeming hopelessly pedantic, I would like to characterize the syntax of our language a bit more carefully. We begin with an odd sort of definition. We shall simultaneously define the *terms* and *formulas* of our language using each notion in the definition of the other. We are rescued from circularity by the fact that in defining the terms or formulas having a certain length n we need only refer to formulas or terms whose length is less than n. So our definitions can be regarded as rules for forming longer and longer expressions out of shorter ones.

Definition: Every variable and individual constant is a term. If φ is a formula and α is a variable, then $\{\alpha: \varphi\}$ is a term.

Definition: If β and γ are terms, then $\beta \in \gamma$ and $\beta = \gamma$ are formulas. If φ is a formula and α is a variable, then $\forall \alpha \varphi$ is a formula. If φ and ψ are both formulas, then so are $(\varphi \rightarrow \psi)$ and $\neg \varphi$.

For example: 'x' is a variable and 'a' is an individual constant. So '$x=a$' is a formula. So '$\{x: x=a\}$' is a term. 'y' is a variable. So '$y \in \{x: x=a\}$' is a formula. So '$\forall y\ y \in \{x: x=a\}$' is a formula. So '$\neg \forall y\ y \in \{x: x=a\}$' is a formula. So '$(x=a \rightarrow \neg \forall y\ y \in \{x: x=a\})$' is a formula. And so on.

Definition: An occurrence of a variable α is said to be *bound* when it is within a formula of the form $\forall \alpha \varphi$ or $\{\alpha: \varphi\}$. Unbound occurrences of a variable are said to be *free*.

For example: There are three occurrences of the variable 'x' in the formula '$(x=a \rightarrow \neg \forall y\ y \in \{x: x=a\})$'. The first occurrence is free. The last two are bound (because they figure within a set operator whose initial variable is 'x').

Definition: A *sentence* is a formula in which no variable occurs free.

For example: '$(x=a \rightarrow \neg \forall y\ y \in \{x: x=a\})$' is not a sentence, but '$\forall x (x=a \rightarrow \neg \forall y\ y \in \{x: x=a\})$' is.

Definition: Let φ be a formula, α a variable, and β either an individual constant or a set operator. Then $\varphi \alpha / \beta$ is the result of replacing each free occurrence of α in φ by an occurrence of β.

For example, if φ is '$(x=a \rightarrow \neg \forall y\ y \in \{x: x=a\})$', α is 'x', and β is 'b', then $\varphi \alpha / \beta$ is '$(b=a \rightarrow \neg \forall y\ y \in \{x: x=a\})$'.

We are now in a position to learn a formal system of deduction. A *formal proof* is a sequence of lines each consisting of a line number, a sentence, an annotation, and, in many cases, one or more premise numbers. For example:

Premise #'s	Line #'s	Sentences	Annotation
1	(1)	$(\forall x\ x \in x \rightarrow \forall y\ y = y)$	A
2	(2)	$(\forall x\ x \in x \rightarrow \neg \forall y\ y = y)$	A
1,2	(3)	$\neg \forall x\ x \in x$	1,2 $\neg$ I

The annotation on each line cites the *rule of inference* we have used to obtain the sentence on that line. It also lists the line numbers of the sentences to which we have applied the rule (if there are any such sentences). The annotation on line 3 says that we obtained '$\neg \forall x\ x \in x$' by applying the rule of negation introduction (abbreviated '$\neg$ I') to the sentences on lines 1 and 2. Our derivation has not established the truth of '$\neg \forall x\ x \in x$' – it has established merely that '$\neg \forall x\ x \in x$' is derivable from the sentences on lines 1 and 2. We make a record of this deductive relation by writing '1,2' in the premise number column on line 3. The premise numbers on the first two lines play a somewhat different role: they indicate that the sentences on those lines were merely assumed rather than derived. (Rule A is our rule of assumption.) All of this should make a bit more sense when we have familiarized ourselves with the rules of inference.

Suppose φ, ψ are formulas, α is a variable, and β, γ are terms in which no variable occurs free.

Rule A (Assumption): Any sentence may appear on a line as long as the premise number on that line is the same as the line number.

Rule $\neg$ I (Negation Introduction): $\neg \varphi$ may be written on a line n if $(\varphi \rightarrow \psi)$ and $(\varphi \rightarrow \neg \psi)$ appear on earlier lines m and m'; write all the premise numbers from m and m' in the premise number column on n.

Rule $\neg$ E (Negation Elimination): φ may be written on a line n if $\neg \neg \varphi$ appears on an earlier line m; write all the premise numbers from m in the premise number column on n.

Rule $\rightarrow$ I (Conditional Introduction): $(\varphi \rightarrow \psi)$ may be written on a line n if ψ appears on an earlier line m; write all the premise numbers from m in the premise number column on n omitting at most those numbers which represent φ. (A premise number p is said to represent a sentence φ if φ occupies the sentence column on line p.)

Rule $\forall$ I (Universal Introduction): Suppose β is an individual constant which does not occur in φ. $\forall \alpha \varphi$ may be written on a line n if $\varphi \alpha / \beta$ appears on an earlier line m and no premise number on m represents a sentence in which β occurs; write all the premise numbers from m in the premise number column on n.

Rule $\forall$ E (Universal Elimination): $\varphi \alpha / \beta$ may be written on a line n if $\forall \alpha \varphi$ appears on an earlier line m; write all the premise numbers from m in the premise number column on n.

Rule $= I$ (Identity Introduction): $\beta = \beta$ may be written on any line; leave the premise number column on that line empty.

Rule $= E$ (Identity Elimination): Suppose ψ is the result of replacing one or more occurrences of β in φ by occurrences of γ. ψ may be written on a line n if φ and $\beta = \gamma$ appear on earlier lines m and m'; write all the premise numbers from m and m' in the premise number column on n.

All of these rules are employed in the following proof that '$\forall x \, x \in b$' is derivable from '$\forall x(x = x \to x \in a)$' and '$a = b$'.

1	(1)	$\forall x(x = x \to x \in a)$	A
2	(2)	$a = b$	A
1	(3)	$(c = c \to c \in a)$	1 $\forall$E
4	(4)	$\neg c \in a$	A
4	(5)	$(c = c \to \neg c \in a)$	4 $\to$I
1,4	(6)	$\neg c = c$	3,5 $\neg$I
1	(7)	$(\neg c \in a \to \neg c = c)$	6 $\to$I
	(8)	$c = c$	=I
	(9)	$(\neg c \in a \to c = c)$	8 $\to$I
1	(10)	$\neg \neg c \in a$	7,9 $\neg$I
1	(11)	$c \in a$	10 $\neg$E
1	(12)	$\forall x \, x \in a$	11 $\forall$I
1,2	(13)	$\forall x \, x \in b$	2,12 =E

Since I don't expect my readers actually to construct any formal derivations, I won't offer guidelines for doing so. It would be a useful exercise, however, for the reader to examine the above proof and verify that on each line the rule cited in the annotation has been correctly applied. In what follows, I shall assume that the reader knows our rules well enough to have completed this exercise.

2. The Principle of Comprehension

The rules we have introduced so far govern only the logical symbols '$\neg$', '$\to$', '$\forall$', and '$=$'. The time has come to introduce rules which deal more directly with the set theoretic vocabulary (i.e., epsilon and the set operators). The following two rules are marked with '†' to indicate that we are adopting them only provisionally.

Rule $\in I$† (Epsilon Introduction): $\beta \in \{\alpha : \varphi\}$ may be written on a line n if $\varphi \alpha / \beta$ appears on an earlier line m; write all the premise numbers from m in the premise number column on n.

86 Some Metamathematics

Rule ∈*E*† (Epsilon Elimination): φα/β may be written on a line *n* if β∈{α: φ} appears on an earlier line *m*; write all the premise numbers from *m* in the premise number column on *n*.

The following derivation scheme will show that we are wise to handle these rules gingerly. By employing the schematic variable 'φ', we establish that a derivation of the following sort can be carried out no matter what sentence is substituted for 'φ'.

1	(1)	{x: ¬x∈x}∈{x: ¬x∈x}	A
1	(2)	¬{x: ¬x∈x}∈{x: ¬x∈x}	1 ∈E†
	(3)	({x: ¬x∈x}∈{x: ¬x∈x} → ¬{x: ¬x∈x}∈{x: ¬x∈x})	2 →I
	(4)	({x: ¬x∈x}∈{x: ¬x∈x} → {x: ¬x∈x}∈{x: ¬x∈x})	1 →I
	(5)	¬{x: ¬x∈x}∈{x: ¬x∈x}	3,4 ¬I
	(6)	{x: ¬x∈x}∈{x: ¬x∈x}	5 ∈I†
	(7)	(¬φ → ¬{x: ¬x∈x}∈{x: ¬x∈x})	5 →I
	(8)	(¬φ → {x: ¬x∈x}∈{x: ¬x∈x})	6 →I
	(9)	¬¬φ	8 ¬I
	(10)	φ	9 ¬E

If a line of a properly constructed derivation has only a blank space in the premise number column, the sentence on that line is said to be a *theorem* of our system. The above proof scheme establishes that *every* sentence is a theorem of the system which includes ∈I† and ∈E†. Deductive systems which have this distressing property (known as "formal inconsistency") are utterly devoid of mathematical interest. They are conservative extensions only of other formally inconsistent systems. And they are likewise interpretable only in their inconsistent brethren. On the other hand, since *every* deductive system is interpretable in a formally inconsistent theory, such interpretations cannot be used to make interesting discriminations between systems. The moral of this story is that either ∈I† or ∈E† or both must be altered in such a way that derivations of the above sort cannot be carried out. Some informal reflections will help us to decide how this can best be done.

At a more or less commonsense level, formulas of the form

$$\varphi\alpha/\beta \leftrightarrow \beta \in \{\alpha: \varphi\}$$

appear superficially to be uncontroversial truths. For example, the formula

$$\forall y \forall z (y = z \leftrightarrow y \in \{x: x = z\})$$

just says that y is identical to z if and only if y is a member of the set of things identical to z. And this seems entirely reasonable (as long as we are willing to accept the existence of sets). Unfortunately, however, if we adopted every formula having this apparently reasonable form, our system would be inconsistent. For from

$$\forall y(y \notin y \leftrightarrow y \in \{x: x \notin x\})$$

(which just says that something is not a member of itself if and only if it is a member of the set of things which are not members of themselves) we can infer

$$\{x: x \notin x\} \notin \{x: x \notin x\} \leftrightarrow \{x: x \notin x\} \in \{x: x \notin x\}$$

which is a contradiction. Although this result is known as "Russell's Antinomy," it was originally discovered by Ernst Zermelo.[80] The formula from which the contradiction arises ('$\forall y(y \notin y \leftrightarrow y \in \{x: x \notin x\})$') is a special case of the logically false formula

$$\forall y(\neg Ryy \leftrightarrow Rya).$$

Russell's Antinomy is just one of the infinitely many "paradoxes of set theory." Two other particularly famous ones are Georg Cantor's paradox of the greatest cardinal and Cesare Burali-Forti's paradox of the greatest ordinal.[81] If our system is to be consistent, we must steer clear of all these hazards.

The scheme '$\varphi\alpha/\beta \leftrightarrow \beta \in \{\alpha: \varphi\}$' is often called the "'principle of comprehension" or the "comprehension scheme" (or even just "Comprehension" for short). Russell's Antinomy shows us that we must place some sort of restriction on the formulas which are used to fill out the comprehension scheme (that is, the formulas which are put in place of 'φ'). But what should these restrictions be? There are so many different set theories available today largely because logicians have dreamed up so many different ways of answering this question – that is, they have dreamed up so many different ways of restricting the principle of comprehension. Which approach should we choose?

Some fairly natural restrictions present themselves when we notice that Comprehension is a biconditional whose constituent conditionals can be considered separately. (The reason for treating the constituent conditionals separately is simply that they make significantly different claims and that these differences ought to be explicitly acknowledged.) Let us first examine the right to left conditional:

$$\beta \in \{\alpha: \varphi\} \rightarrow \varphi\alpha/\beta.$$

This scheme just says that if β is a member of the set of α which satisfy the predicate φ, then β satisfies φ. In other words, if the set of things which satisfy φ has any members at all, those members satisfy φ. This principle places a restriction on the objects which can be members of a set. So it is a conservative, limitative principle which does not itself seem to require restriction. Accordingly, we seem to be justified in placing it within our system without any amendment. This amounts to accepting ∈E† exactly as it is formulated above. We shall do so and shall now refer to it simply as "∈E" (without the invidious '†'). This allows us to prove that there is a set which has no members.

Theorem 1 $\forall y \neg y \in \{x: \neg x = x\}$.

1	(1)	$a \in \{x: \neg x = x\}$	A
1	(2)	$\neg a = a$	1 ∈E
	(3)	$(a \in \{x: \neg x = x\} \rightarrow \neg a = a)$	2 →I
	(4)	$a = a$	=I
	(5)	$(a \in \{x: \neg x = x\} \rightarrow a = a)$	4 →I
	(6)	$\neg a \in \{x: \neg x = x\}$	3,5 ¬I
	(7)	$\forall y \neg y \in \{x: \neg x = x\}$	6 ∀I

A set which has no members is said to be empty. It is of some interest that the existence of an empty set follows from the unrestricted use of a principle stating that such-and-such a set has *no more than* such-and-such members, whereas (as we shall see) the existence of non-empty sets follows from a restricted use of a principle stating that such-and-such a set has *at least* such-and-such members. So empty and non-empty sets differ with respect to the grounds on which they might be taken to exist.[82]

Now let us consider the principle of comprehension taken left to right:

$$\varphi\alpha/\beta \rightarrow \beta \in \{\alpha: \varphi\}.$$

This scheme says that if an object satisfies a given predicate, then it is a member of a set consisting of, at least, all the objects which satisfy that predicate. This principle implies the existence of sets which play a role in the generation of set theoretic paradoxes: for example, the set consisting of, at least, all the sets which are not members of themselves; or the set consisting of, at least, all the ordinals; or the set consisting of, at least, all the cardinals; or even the set consisting of, at least, everything in the universe. So anyone familiar with Russell's Antinomy and its ilk would feel an immediate urge to restrict this scheme in some way or to exercise some other sort of damage control. We shall take an approach which is essentially due to Ernst Zermelo: namely, to adopt those instances of this scheme which serve our immediate purposes and which seem not to lead to contradictions.[83] For the moment, we shall, in effect, adopt only the following instance:

$$\forall x \forall y(x=y \rightarrow x \in \{x: x=y\}).$$

Translated into a rule of our deductive system, we have:

Rule $\in I$: $\beta \in \{\alpha: \alpha=\gamma\}$ may be written on a line n if $\beta=\gamma$ appears on an earlier line m; write all the premise numbers from m in the premise number column on n.

As promised, we are now in a position to prove that there are non-empty sets.

Theorem 2 $a \in \{x: x=a\}$
(1) $a=a$ $=I$
(2) $a \in \{x: x=a\}$ $1 \in I$

As it now stands, our deductive system is extremely weak. And it certainly doesn't count as a universal theory. Nonetheless, we shall be able to put it to good use in our study of interpretability. Let's call our system "Σ."

3. Interpretation of the Successor Function

I would like to show how a weak theory of the successor function can be interpreted within Σ. The logical symbols of our new theory as well as the rules of inference governing those symbols will be the same as those of Σ. The non-logical vocabulary will contain the unary operation symbol 'S' and the individual constant '0'. The terms of our new language are characterized as follows:

Definition: '0' is a term; every variable and individual constant is a term; and if β is a term, then so is $S(\beta)$.

Given any natural number n, $S(n)$ is simply $n+1$.

We adopt just two new rules:

Rule SI (Successor Introduction): $\neg 0=S(\beta)$ may be written on any line; leave the premise number column on that line empty.

Rule SE (Successor Elimination): $\beta=\gamma$ may be written on a line n if $S(\beta)=S(\gamma)$ appears on an earlier line m; write all the premise numbers from m in the premise number column on n.

The point behind SI is that 0 is not the successor of any natural number. The point behind SE is that natural numbers with identical successors are themselves identical.

Let's call our new system (resulting from the deletion of $\in I/\in E$

90 SOME METAMATHEMATICS

and the addition of SI/SE) "Σ_ω." We shall now define an interpretation function i which maps theorems of Σ_ω into theorems of Σ. i will deal with the sentential connectives '¬' and '→' in the way described in chapter 4. And i can handle the quantifier '∀' in the first way we described:

$$i(\forall\alpha\varphi) \text{ is } \forall\alpha i(\varphi).$$

We saw that it would be overly restrictive to require all of our interpretations to have this form. But, in the present case, this approach is fine. Alternatively, if we wanted all our interpretation functions to handle the quantifiers in the same basic way, then, in the present case, we could say that

$$i(\forall\alpha\varphi) \text{ is } \forall\alpha(\alpha=\alpha \to i(\varphi)).$$

Clearly, the predicate $\alpha=\alpha$ satisfies the requirement that $\exists\alpha\ \alpha=\alpha$ (i.e., $\neg\forall\neg\alpha=\alpha$) be a theorem of Σ. (This requirement was introduced back in our original discussion of interpretability.) And one can see that this way of dealing with '∀' will give essentially the same results as the one first proposed (since the predicate $\alpha=\alpha$ is vacuous – that is, it is automatically satisifed by everything in our universe). Since we have no particular reason to insist upon the absolute uniformity of our interpretation functions, we shall use the simpler version of i (without the clause $\alpha=\alpha$).

We must now explain what i is to do with '=', 'S', '0', the variables, and the individual constants. If β,γ are terms, δ is an individual constant, and α is a variable which is distinct both from β and from any variable which occurs in β, then

$$\begin{aligned} i(\beta=\gamma) &\text{ is } i(\beta)=i(\gamma) \\ i(S(\beta)) &\text{ is } \{\alpha: \alpha=i(\beta)\} \\ i(0) &\text{ is } \{x: \neg x=x\} \\ i(\alpha) &\text{ is } \alpha \\ i(\delta) &\text{ is } \delta. \end{aligned}$$

We require that α be distinct from β and from any variable which occurs in β because, for example, we want to prevent 'S(x)' from being mapped into '{x: x=x}'. For we want $i(S(x))$ to be the set whose only member is x, not the set of all self-identical objects. In order to characterize i more precisely, we might require that $i(S(\beta))$ be $\{x: x=\beta\}$ whenever 'x' neither is nor occurs in β and that $i(S(\beta))$ be $\{y: y=\beta\}$ otherwise. (Note that no more than one variable can occur in a term of the system Σ_ω. So if 'x' occurs in β, then 'y' does not.)

The idea behind our interpretation function (due essentially to Ernst Zermelo) is that zero can be taken to be the empty set, while the successor function can be taken to be the operation of forming a unit set.[84] According to this conception, the successor of zero is the set whose only member is the empty set. The successor of the successor of zero is the set whose only member is the set whose only member is the empty set. And so on.

We must now prove that $i(\varphi)$ is a theorem of Σ whenever φ is a theorem of Σ_ω. In fact, we shall prove the more general result that $i(\psi)$ is derivable from $i(\varphi_1), \ldots, i(\varphi_m)$ in Σ whenever ψ is derivable from $\varphi_1, \ldots, \varphi_m$ in Σ_ω. The idea is to show that every move in a derivation of ψ can be mimicked to produce a derivation of $i(\psi)$.

1. Pick some particular derivation of ψ from $\varphi_1, \ldots, \varphi_m$ in Σ_ω. In this Σ_ω-derivation of ψ, we use Rule A to write down $\varphi_1, \ldots, \varphi_m$ and, possibly, some additional assumptions $\theta_1, \ldots, \theta_k$. In the corresponding Σ-derivation of $i(\varphi)$, we can equally well use Rule A to write down $i(\varphi_1), \ldots, i(\varphi_m)$ and the additional assumptions $i(\theta_1), \ldots, i(\theta_k)$. Suppose (without loss of generality) that all applications of Rule A occur in the initial $m+k$ lines of the Σ_ω-derivation. We shall now consider what happens on line $m+k+1$.

2. Suppose that $(\varphi \to \theta)$ and $(\varphi \to \neg \theta)$ appear among the assumptions of the Σ_ω-derivation and that we now apply $\neg$I to get $\neg \varphi$. $i((\varphi \to \theta))$ and $i((\varphi \to \neg \theta))$ will appear among the assumptions of the Σ-derivation. And, by the definition of i, these formulas are just $(i(\varphi) \to i(\theta))$ and $(i(\varphi) \to \neg i(\theta))$. So we can apply $\neg$I to get $\neg i(\varphi)$ – that is, $i(\neg \varphi)$.

3. Suppose that $\neg \neg \varphi$ appears among the assumptions of the Σ_ω-derivation and that we now apply $\neg$E to get φ. $i(\neg \neg \varphi)$ will appear among the assumptions of the Σ-derivation. And, by the definition of i, this formula is just $\neg \neg i(\varphi)$. So we can apply $\neg$E to get $i(\varphi)$.

4. Suppose that θ appears among the assumptions of the Σ_ω-derivation and that we now apply $\to$I to get $(\varphi \to \theta)$. $i(\theta)$ will appear among the assumptions of the Σ-derivation. So we can apply $\to$I to get $(i(\varphi) \to i(\theta))$ – that is, $i((\varphi \to \theta))$. Suppose that, in the Σ_ω-derivation, we delete a premise number when we apply $\to$I. Then, since we have obtained θ by Rule A, this premise number must represent θ itself. In the Σ-derivation, we can similarly delete a premise number which represents $i(\theta)$.

5. Suppose that $\varphi\alpha/\beta$ appears among the assumptions of the Σ_ω-derivation and that no premise number written to the left of $\varphi\alpha/\beta$ represents a formula in which the individual constant β occurs. Since we have used Rule A to obtain $\varphi\alpha/\beta$, this formula itself will be represented by the single premise number written to its left. So $\varphi\alpha/\beta$ must contain no occurrences of β (which is possible as long as φ contains no free occurrences of α). This just means that $\varphi\alpha/\beta$ is φ itself.

Suppose, now, that we apply ∀I to get ∀αφ. $i(\varphi\alpha/\beta)$ – that is, $i(\varphi)$ – will appear among the assumptions of the Σ-derivation. Just as in the Σ_ω-derivation, we can apply ∀I to get ∀α$i(\varphi)$ – that is, $i(\forall\alpha\varphi)$.

6. Suppose that ∀αφ appears among the assumptions of the Σ_ω-derivation and that we now apply ∀E to get φα/β. Remember that β must contain no free occurrences of variables. $i(\forall\alpha\varphi)$ will appear among the assumptions of the Σ-derivation. And, by the definition of i, this formula is just ∀α$i(\varphi)$. Furthermore, by the definition of i, $i(\beta)$ will contain no free occurrences of variables – and $i(\varphi)$ will contain free occurrences of α in positions corresponding to the free occurrences of α in φ. So we can apply ∀E to get $i(\varphi\alpha/\beta)$.

7. Suppose that we use =I to write β=β on line $m+k+1$ of the Σ_ω-derivation. Then we can use =I to write $i(\beta)=i(\beta)$ – that is, $i(\beta=\beta)$ – on line $m+k+1$ of the Σ-derivation.

8. Suppose that φ and β=γ appear among the assumptions of the Σ_ω-derivation and that we now apply =E to get φ′ – where φ′ is the result of replacing one or more occurrences of β in φ by occurrences of γ. Remember that β and γ must contain no free occurrences of variables. $i(\varphi)$ and $i(\beta=\gamma)$ will appear among the assumptions of the Σ-derivation. And, by the definition of i, the latter formula is just $i(\beta)=i(\gamma)$. Furthermore, by the definition of i, neither $i(\beta)$ nor $i(\gamma)$ will contain free occurrences of variables – and $i(\varphi)$ will contain occurrences of $i(\beta)$ in positions corresponding to the occurrences of β in φ. So we can apply =E to get $i(\varphi')$.

9. Suppose that we use SI to write ¬0=S(β) on line $m+k+1$ of the Σ_ω-derivation. By the definition of i, $i(\neg 0 = S(\beta))$ is ¬{x: ¬$x=x$}={α: α=$i(\beta)$} – which we can introduce into the Σ-derivation by means of a sub-derivation having the following form. For the sake of readability, we shall allow ourselves to negate identity statements by writing '≠' in place of '='. For example, '$x \neq x$' is to have the same meaning as '¬$x=x$'.

j	(j)	{x:$x \neq x$}={α:α=$i(\beta)$}	A
	$(j+1)$	$i(\beta)=i(\beta)$	=I
	$(j+2)$	$i(\beta) \in$ {α:α=$i(\beta)$}	$j+1 \in$I
	$(j+3)$	{x:$x \neq x$}={x:$x \neq x$}	=I
j	$(j+4)$	{α:α=$i(\beta)$}={x:$x \neq x$}	$j, j+3$ =E
j	$(j+5)$	$i(\beta) \in$ {x:$x \neq x$}	$j+2, j+4$ =E
j	$(j+6)$	$i(\beta) \neq i(\beta)$	$j+5$ ∈E
	$(j+7)$	({x:$x \neq x$}={α:α=$i(\beta)$} → $i(\beta) \neq i(\beta)$)	$j+6$ →I
	$(j+8)$	({x:$x \neq x$}={α:α=$i(\beta)$} → $i(\beta)=i(\beta)$)	$j+1$ →I
	$(j+9)$	{x:$x \neq x$} ≠ {α:α=$i(\beta)$}	$j+7, j+8$ ¬I

INTERPRETATION OF THE SUCCESSOR FUNCTION 93

10. Suppose that $S(\beta)=S(\gamma)$ appears among the assumptions of the Σ_ω-derivation and that we now apply SE to get $\beta=\gamma$. $i(S(\beta)=S(\gamma))$ will appear among the assumptions of the Σ-derivation. And, by the definition of i, this formula is just $\{\alpha\colon \alpha=i(\beta)\}=\{\alpha\colon \alpha=i(\gamma)\}$. So we can use a sub-derivation of the following form to get $i(\beta)=i(\gamma)$ – that is, $i(\beta=\gamma)$.

j	(j)	$\{\alpha\colon \alpha=i(\beta)\}=\{\alpha\colon \alpha=i(\gamma)\}$	A
	$(j+1)$	$i(\beta)=i(\beta)$	$=$I
	$(j+2)$	$i(\beta)\in\{\alpha\colon\alpha=i(\beta)\}$	$j+1\in$I
j	$(j+3)$	$i(\beta)\in\{\alpha\colon \alpha=i(\gamma)\}$	$j, j+2 =$E
j	$(j+4)$	$i(\beta)=i(\gamma)$	$j+3 \in$E

Suppose that line $m+k+1$ of the Σ_ω-derivation features a sentence λ and premise numbers representing sentences $\lambda_1,...,\lambda_j$. Then the above construction guarantees that a line of our Σ-derivation will feature the sentence $i(\lambda)$ and premise numbers representing the sentences $i(\lambda_1),...,i(\lambda_j)$. We must now establish that the contents of line $m+k+2$ of the Σ_ω-derivation can be mimicked in this same way within the Σ-derivation. As the reader might already have guessed, the above construction can be repeated at this point with only trivial amendments. For example, Clause 4 should be rewritten as follows.

4'. Suppose that θ appears among the first $m+k+1$ lines of the Σ_ω-derivation and that we now apply →I to get $(\varphi\to\theta)$. $i(\theta)$ will appear within the Σ-derivation. So we can apply →I to get $(i(\varphi)\to i(\theta))$ – that is, $i((\varphi\to\theta))$. Suppose that, in the Σ_ω-derivation, we delete a premise number when we apply →I. If we obtained θ by Rule A, this premise number must represent θ itself. In the Σ-derivation, we can similarly delete a premise number which represents $i(\theta)$. If θ occupies line $m+k+1$, the premise number we delete will represent one of $\lambda_1,\ldots,\lambda_j$. In the Σ-derivation, we can similarly delete a premise number which represents one of $i(\lambda_1),\ldots,i(\lambda_j)$.

To Clause 5 we must add the following codicil.

5'. Suppose that $\varphi\alpha/\beta$ appears on line $m+k+1$ of the Σ_ω-derivation, that β is an individual constant which does not occur in either φ or $\lambda_1,\ldots,\lambda_j$, and that we apply ∀I on line $m+k+2$ to get $\forall\alpha\varphi$. $i(\varphi\alpha/\beta)$ will appear on a line of the Σ-derivation. By the definition of i, $i(\beta)$ (which is just β itself) does not occur in either $i(\varphi)$ or $i(\lambda_1),\ldots,i(\lambda_j)$. So we may apply ∀I to get $\forall\alpha i(\varphi)$ – that is, $i(\forall\alpha\varphi)$. (This last step depends on the fact that $i(\varphi\alpha/\beta)$ is the result of replacing every free occurrence of α in $i(\varphi)$ by an occurrence of β. In other words, $i(\varphi\alpha/\beta)$ is just $i(\varphi)\alpha/\beta$.)

Suppose that line $m+k+2$ of the Σ_ω-derivation features a sentence λ' and premise numbers representing sentences $\lambda'_1,\ldots,\lambda'_j$. Then our

94 Some Metamathematics

revised construction guarantees that a line of our Σ-derivation will feature the sentence $i(\lambda')$ and premise numbers representing the sentences $i(\lambda'_1), \ldots, i(\lambda'_j)$. We can now move on to line $m+k+3$ of the Σ_ω-derivation and, by repeating the above procedure, can mimic its contents on a line of the Σ-derivation. In fact, it should be clear that we can continue this process until we reach the final line of the Σ_ω-derivation – whose contents will then be mimicked on the final line of the Σ-derivation.

Let's see how this works in a particular case. First, consider the following Σ_ω-derivation.

1	(1)	$S(0)=S(S(0))$	A
1	(2)	$0=S(0)$	1 SE
	(3)	$0 \neq S(0)$	SI
	(4)	$(S(0)=S(S(0))) \to 0=S(0))$	2 $\to$I
	(5)	$(S(0)=S(S(0))) \to 0 \neq S(0))$	3 $\to$I
	(6)	$S(0) \neq S(S(0))$	4,5 $\neg$I

Using the steps described above, we transform this into the following Σ-derivation. For the sake of readability, let's abbreviate '$\{x: x \neq x\}$' as '$\varnothing$' and $\{\alpha: \alpha=\beta\}$ as $\{\beta\}$.

1	(1)	$\{\varnothing\}=\{\{\varnothing\}\}$	A
	(2)	$\varnothing=\varnothing$	=I
	(3)	$\varnothing \in \{\varnothing\}$	2 $\in$I
1	(4)	$\varnothing \in \{\{\varnothing\}\}$	1,3, =E
1	(5)	$\varnothing=\{\varnothing\}$	4 $\in$E
6	(6)	$\varnothing=\{\varnothing\}$	A
	(7)	$\varnothing=\varnothing$	=I
	(8)	$\varnothing \in \{\varnothing\}$	7 $\in$I
	(9)	$\varnothing=\varnothing$	=I
6	(10)	$\{\varnothing\}=\varnothing$	6,9 =E
6	(11)	$\varnothing \in \varnothing$	8,10 =E
6	(12)	$\varnothing \neq \varnothing$	11 $\in$E
	(13)	$(\varnothing=\{\varnothing\} \to \varnothing \neq \varnothing)$	12 $\to$I
	(14)	$(\varnothing=\{\varnothing\} \to \varnothing=\varnothing)$	7 $\to$I
	(15)	$\varnothing \neq \{\varnothing\}$	13,14, $\neg$I
	(16)	$(\{\varnothing\}=\{\{\varnothing\}\} \to \varnothing=\{\varnothing\})$	5 $\to$I
	(17)	$(\{\varnothing\}=\{\{\varnothing\}\} \to \varnothing \neq \{\varnothing\})$	15 $\to$I
	(18)	$\{\varnothing\} \neq \{\{\varnothing\}\}$	16,17 $\neg$I

Line 1 of this Σ-derivation corresponds to line 1 of the Σ_ω-derivation, lines 2–5 correspond to line 2, lines 6–15 correspond to line 3, and lines 16–18 correspond to lines 4–6. The Σ-derivation is terribly inelegant. (We could shorten it to eleven lines if we wished.) But then I never promised that our construction procedure would always generate proofs which are as brief as possible.

Our two derivations reveal how a proof within arithmetic that S(0) is not identical to S(S(0)) can be transformed, without any loss of its essential logical features, into a proof within set theory that $\{\varnothing\}$ is not identical to $\{\{\varnothing\}\}$. This is precisely the sort of thing I had in mind when I remarked, in chapter 1, that the patterns of inference characteristic of all but the most heterodox mathematical sub-fields can be reproduced within widely accepted set theories. More generally, our study of interpretation functions has placed within our reach a technically informed account of set theory's "foundational" status. We say that, for example, the set theory ZF can be considered foundational with respect to an informal web of mathematical practices Π if and only if the essential features of Π (as judged by its practitioners) can be captured within a formal deductive system Π' and, furthermore, Π' is interpretable in ZF. (If, as often happens, Π' is simply ZF itself, we needn't worry ourselves about interpretability.) I have no general account of what it means to "capture the essential features" of a collection of informal practices. This is a matter which should probably be left up to the mathematicians who engage in those practices.[85] But once a group of specialists has given their blessing to a formalization Π', it is open to anyone versed in set theory and logic to determine whether Π' is interpretable in, say, ZF. As chapter 1 suggested, this process of formalization and interpretation has been applied to all the informal practices which make up the mathematical mainstream — or, at least, it is generally taken for granted that it *could* be so applied. This is a crucial part of set theory's claim to the title of *the* foundation for mathematics.

Our study of Σ and Σ_ω has involved us in precisely the sort of behavior which formalists regard as central to mathematics. To a formalist, the purest and most highly evolved forms of mathematical activity are the creation of formal deductive systems, the derivation of formulas within those systems, and the investigation of the properties of such systems. True, flesh and blood mathematicians almost never construct formal derivations, relying instead on highly compressed, informal demonstrations. But the untidiness of actual mathematical practices need not trouble the formalist — as long as those practices can be regarded as raw material for the construction of formal theories and derivations. A practicing mathematician's unwavering attachment to informal proofs need not cause alarm: informal proofs can, after all, be regarded as valuable pieces of evidence for

96 SOME METAMATHEMATICS

the existence of formal ones and, so, bring excellent grist to the formalist mill.

This completes our brief foray into metamathematics. After a transitional chapter, we shall begin to examine an approach to the philosophical interpretation of set theory which is largely incompatible with the formalist outlook discussed here and in chapter 5. We shall move from formalism to mathematical structuralism.

VII

LOGICAL INTERLUDE: SECOND ORDER LOGIC

1. Logics of the First and Higher Orders

To prepare the way for our remaining chapters, we must once again familiarize ourselves with some basic logical notions. Our particular concern at the moment is second order logic. First, second, and higher order logics contain expressions called "variables" which (as Willard Van Orman Quine has emphasized) play roughly the same grammatical and semantic role as the pronouns ('he', 'she', 'it', . . .) of ordinary English. These logics also contain inference rules which allow the interchange of variables with other sorts of expressions. In first order logics, variables can be interchanged only with expressions which function grammatically and semantically like nouns or pronouns. In higher order logics, variables can be interchanged with expressions which function like predicates.

We need to be clear about what exactly is meant by a predicate here. Consider the English declarative sentence

Ruskin is under the pediment.

If we replace the word 'Ruskin' by a blank, we obtain a unary predicate:

——— is under the pediment.

If we replace the words 'the pediment' by a blank, we obtain another unary predicate:

Ruskin is under ———.

And if we replace both 'Ruskin' and 'the pediment' by blanks, we obtain a binary predicate:

——— is under ———.

Whenever we put blanks in place of nouns or noun-like expressions in a declarative sentence, we obtain a (unary, binary, ternary, . . .)

98 Logical Interlude: Second Order Logic

predicate. As I just indicated, the order of a logic is determined, at least in part, by the expressions which are allowed to trade places with variables. Logics which let predicates play this game are of an order higher than the first.

Examples of first order inferences are in no short supply. A particularly familiar form of first order inference (corresponding to our ∀E) takes us from the assertion that everything in the world has such-and-such a property to the assertion that some particular thing has that property. For example, if (absurdly) we accepted the proposition that everything in the world has congregated under a certain pediment, then a first order logic would allow us to infer that some particular object or person — say, Ruskin — is under that pediment. If we allow ourselves to express this inference in rather stilted language:

> Everything is such that it is under the pediment.
> Therefore, Ruskin is under the pediment.

we can detect the characteristic interchange of pronoun ('it') and noun-like expression ('Ruskin') — accompanied here by the deletion of the quantifier 'Everything is such that'.

In order to illustrate the corresponding sort of second order inference, we need to add some new elements to our ordinary English — in particular, special pronouns 'IT^1< >', 'IT^2<,>', 'IT^3<,,>', ... which are to be replaced not by noun-like expressions, but by predicates. With these new linguistic tools in hand, we can perform a second order universal instantiation:

> Everything is such that IT^1<Ruskin>.
> Therefore, Ruskin is under the pediment.

As before, we delete the quantifier 'Everything is such that'. But now we replace the "pronoun" 'IT^1< >' by the unary *predicate* '——— is under the pediment', letting 'Ruskin' fill the blank. This is a second order inference. Another example:

> Everything is such that IT^2<Ruskin, the pediment>.
> Therefore, Ruskin is under the pediment.

As usual, we drop 'Everything is such that'. But here we replace 'IT^2<,>' by the binary predicate '——— is under ———', letting 'Ruskin' and 'the pediment', respectively, fill the blanks. This is also a second order inference.

Of course, 'Everything is such that IT^2<Ruskin, the pediment>' is not English. But some such queer neologism seems necessary if we are to avoid reading more into the second order formalism than is there. Many of the closest ordinary English analogues of second order inferences involve explicit references to entities whose very existence is a matter of philosophical debate: sets, properties, and relations. For example (continuing our use of stilted language):

> Every set is such that Ruskin is a member of it.
> Therefore, Ruskin is under the pediment.

or

> Every relation is such that Ruskin stands in it to the pediment. Therefore, Ruskin is under the pediment.

Citing inferences such as these would have allowed us to stay closer to ordinary English. But it would also have given the false (or, at least, questionable) impression that second order logic is a theory of sets or of properties or of relations.

It would be somewhat less questionable to say that second order logic is a theory of sets-or-properties-or-relations. Lest the reader think that this is a distinction without a difference, let me hasten to explain myself. There might be no philosophically respectable way to endorse second order logic while giving a blanket denial of the existence of universals (that is, while claiming that neither sets nor properties nor relations nor any other such entities exist). But there probably are philosophically respectable ways of embracing second order logic while denying the existence of universals *of some particular sort* — say, sets or properties or relations (but not all of them at once). So a logician who built a commitment to sets or a commitment to properties or a commitment to relations into the foundations of second order logic would probably be begging significant philosophical questions. In an effort to avoid this, I have been forced to give examples that are not really part of the English language.

Monadic second order inferences are those involving the interchange of pronominal expressions with *unary* predicates (as in the first example of a second order inference given above). As George Boolos has recently shown, we can cast some light on these inferences without either straying too far from ordinary English or reading overly idiosyncratic philosophical theories into the second order formalism.[86] The trick is to exploit the ordinary English device of plural quantification. Plural quantifiers are expressions such as 'there are' which range over *pluralities* of individuals rather than over individuals taken separately. For example, when we say "Some rabbits surrounded the cat" (or, in our more stilted style, "There are rabbits such that they surrounded the cat") we are asserting that a number or plurality of rabbits, acting in concert, surrounded the cat – which is very different from the nonsensical singular generalization "There is a rabbit such that it surrounded the cat." (This example, by the way, is essentially due to Peter Simons.)

Plural quantifiers supply us with ontologically neutral, ordinary language analogues of monadic second order inferences. For instance, consider:

> Given any number of objects, Ruskin is one of them.
> Therefore, Ruskin is one of the things under the pediment.

This corresponds to the quasi-English second order inference:

> Everything is such that IT^1 < Ruskin >.
> Therefore, Ruskin is under the pediment.

Here the pronominal expression 'IT^1 < >' and the plural pronoun 'them' play essentially the same role. 'IT^1 < >' is replaced by the unary predicate '_____ is under the pediment', while 'them' is replaced by the plural definite description 'the things under the pediment'. But, in the above context, the latter substitution yields the predicate '_____ is one of the things under the pediment'. Since being one of the things under the pediment is hardly to be distinguished from being under the pediment, the similarity of the two inferences is evident.

Unfortunately, this way of explaining second order inferences is severely limited. For, while English plural quantifiers can help to clarify *monadic* second order inferences, they will be of no use when we try to explain inferences involving the interchange of pronominal expressions with, for example, binary predicates. It appears that, unless we are willing to saddle second order logic with a commitment to universals of some definite sort, we need to stray rather far from ordinary English to find informal or semi-formal illustrations of such inferences.

2. A Monadic Second Order Deductive System

It will be well worth our while to examine Boolos' work more closely. But first let me introduce a formal version of monadic second order logic. We modify the formal language of chapter 6 by dropping the set theoretic vocabulary ('∈' and set operators) while adding second order variables ('X', 'Y', 'Z' among others) which will play the role of the made-up pronoun 'IT^1 < >'. Our new language will also contain second order descriptors of the form [α: φ] – where α is a first order variable and φ is any formula. For technical reasons, we also adopt a collection of second order constants ('F', 'G', 'H' among others). These changes oblige us to alter our notion of term and formula.

Definition: Every first order variable ('x', 'y', 'z', etc.) and constant ('a', 'b', 'c', etc.) is a first order term. Every second order variable and constant is a second order term. If φ is a formula and α is a first order variable, then [α: φ] is a second order term.

Definition: If β and δ are, respectively, first and second order terms, then δβ is a formula. If β and γ are first order terms, then β=γ is a formula. If φ is a formula and δ is either a first or second order variable, then ∀δφ is a formula. If φ and ψ are both formulas, then so are (φ→ψ) and ¬φ.

For example: 'a' and 'X' are, respectively, first and second order terms. So 'Xa' is a formula. So '∀X Xa' is a formula. So '¬∀X Xa' is a formula. 'x' and 'F' are, respectively, first and second order terms. So 'Fx' is a formula. So '[x: Fx]' is a second order term. 'y' is a first order variable. So '[x: Fx]y' is a formula. So '([x: Fx]y → ¬∀X Xa)' is a formula. So '∀y([x: Fx]y → ¬∀X Xa)' is a formula.

Definition: An occurrence of a first or second order variable α is said to be bound when it is within a formula of the form ∀αφ or [α: φ]. Unbound occurrences of variables are said to be free.

We now adopt rules A, ¬I, ¬E, →I, ∀I, ∀E, =I, and =E from chapter 6 with only minor modifications. In ∀I and ∀E, α may be either a first or second order variable. In ∀I, β may be either a first or second order constant. In ∀E, β may be any first or second order term in which no variable occurs free. In =I and =E, β and γ must be first order constants.

Here's an example of how our quantifier rules can be applied within our second order language.

1	(1)	Fa	A
	(2)	$(Fa \to Fa)$	1 →I
	(3)	$\forall X(Xa \to Xa)$	2 ∀I
	(4)	$([x: Gx]a \to [x: Gx]a)$	3 ∀E
	(5)	$\forall y([x: Gx]y \to [x: Gx]y)$	4 ∀I

We shall adopt the following rules to govern our descriptors. Let α be a first order variable and let β be a first order constant.

Rule []I: [α:φ]β may be written on a line n if φα/β appears on an earlier line m; write all the premise numbers from m in the premise number column on n.

Rule []E: φα/β may be written on a line n if [α:φ]β appears on an earlier line m; write all the premise numbers from m in the premise number column on n.

Our second order descriptors are meant to function like plural definite descriptions: [α: φ] is to be read as "the objects which satisfy the predicate φ." For example, if 'Gx' means that x is under the pediment, then '[x: Gx]' is to be read as "the objects which are under the pediment." If the first order constant 'a' is taken to designate Ruskin, then '[x: Gx]a' is to be read as "Ruskin is one of the objects under the pediment." Given that Ruskin is one of the objects under the pediment (that is, [x: Gx]a), []E allows us to infer that Ruskin is under the pediment (that is, Ga). Conversely, given that Ruskin is under the pediment (that is, Ga), []I allows us to infer that Ruskin is one of the objects under the pediment (that is, [x: Gx]a). The following derivation shows how the characteristic interchange of predicates and second order variables takes place within our system.

1	(1)	$\forall X\ Xa$	A	
1	(2)	[x: Gx]a	1 $\forall$E	
1	(3)	Ga	2 []E	

$\forall$E and []E working together allow us to drop the second order quantifier '$\forall X$' and replace the second order variable 'X' with the predicate 'G——'.

[]I and []E are clearly reminiscent of $\in$I† and $\in$E†. But we needn't worry about Russell's Paradox cropping up. '$x \notin x$' is, of course, not a formula of our second order language. And, more importantly, its nearest analogues ('$\neg XX$', '$\neg FF$', '$\neg[x: Fx][x: Fx]$') are not formulas either. So we are not even in a position to *state* the sort of sentence which Russell used to generate the celebrated contradiction. The set theoretic paradoxes are closely linked to the iteration of set formation (the formation of sets of sets). Since, in the spirit of chapter 3, our "plural logic" does not allow for the formation of pluralities of pluralities, it does not face the same dangers as a set theory.

Here are two more examples of how our new rules operate. In both of them, we shall use '$\exists X$' as an abbreviation for '$\neg \forall \neg X$'.

Theorem 1 $\exists X \forall y\ Xy$

1	(1)	$\forall X \neg \forall y\ Xy$	A
1	(2)	$\neg \forall y\ [x: x=x]y$	1 $\forall$E
	(3)	$a=a$	=I
	(4)	[x: $x=x$]a	3 []I
	(5)	$\forall y\ [x: x=x]y$	4 $\forall$I
	(6)	$(\forall X \neg \forall y\ Xy \rightarrow \neg \forall y\ [x: x=x]y)$	2 $\rightarrow$I
	(7)	$(\forall X \neg \forall y\ Xy \rightarrow \forall y\ [x: x=x]y)$	5 $\rightarrow$I
	(8)	$\exists X \forall y\ Xy$	6,7 $\neg$I

Theorem 2 $\exists X \forall y \neg Xy$

1	(1)	$\forall X \neg \forall y \neg Xy$	A
1	(2)	$\neg \forall y \neg [x: x \neq x]y$	1 $\forall$E
3	(3)	$[x: x \neq x]a$	A
3	(4)	$a \neq a$	3 []E
	(5)	$([x: x \neq x]a \rightarrow a \neq a)$	4 $\rightarrow$I
	(6)	$a = a$	=I
	(7)	$([x: x \neq x]a \rightarrow a = a)$	6 $\rightarrow$I
	(8)	$\neg [x: x \neq x]a$	5,7 $\neg$I
	(9)	$\forall y \neg [x: x \neq x]y$	8 $\forall$I
	(10)	$(\forall X \neg \forall y \neg Xy \rightarrow \neg \forall y \neg [x: x \neq x]y)$	2 $\rightarrow$I
	(11)	$(\forall X \neg \forall y \neg Xy \rightarrow \forall y \neg [x: x \neq x]y)$	9 $\rightarrow$I
	(12)	$\exists X \forall y \neg Xy$	10,11 $\neg$I

We intend to treat '$\exists X$' (that is, '$\neg \forall X \neg$') as a plural existential quantifier. But, as we shall see, doing so will require a few tricks. We would like to read $\exists X \ldots X \ldots$ simply as "There are objects such that ... they/them ..." And, in the case of Theorem 1, this leads to no problems. '$\exists X \forall y Xy$' becomes "There are objects such that everything is one of them." The latter assertion is clearly true – because, for example, everything is one of the objects which are identical to themselves. Theorem 2 is more problematic. If we give '$\exists X \forall y \neg Xy$' the straightforward reading, we get "There are objects such that nothing is one of them" – which is like saying that there are objects which are not any objects in particular. This makes no more sense than saying that, for example, there can be coins in an empty pocket. Whereas '$\exists X \forall y \neg Xy$' is a theorem of our "plural logic," the sentence 'There are objects such that nothing is one of them' seems to be a blatant falsehood. Clearly, our second order existential quantifiers need to be interpreted more subtly. The required subtlety is supplied by George Boolos.

3. Boolos on Monadic Second Order Logic

The problem here is that monadic second order variables act as if they range over empty as well as non-empty "pluralities." Since the notion of a multiplicity of nothing seems absurd, we must discover a trick which renders sentences such as '$\exists X \forall y \neg Xy$' true, but does not commit us to the existence of empty pluralities. Boolos' idea is to read $\exists X \varphi$ as "Either P or Q" – where P is a straightforward translation of $\exists X \varphi$ and Q is a translation of φ which renders $X\alpha$ as $\alpha \neq \alpha$.

Q is meant to handle those cases where 'X' behaves as if it picks out an "empty plurality." If (to speak loosely) X is empty, then $X\alpha$ is false no matter what α is. And, of course, $\alpha \neq \alpha$ is also false no matter what α is. So, when X is empty, $\alpha \neq \alpha$ is a reasonable substitute for $X\alpha$.

Let $\exists X\varphi$ be, for example, '$\exists X \forall y \neg Xy$'. Then P is the sentence 'there are objects such that nothing is one of them'. (Since this is false, we have to hope that Q is true.) If we replace each occurrence of 'Xy' in φ with an occurrence of '$y \neq y$', we get '$\forall y \neg y \neq y$'. And if we translate this into English in a straightforward way, we get 'nothing fails to be self-identical' or, more elegantly, "everything is self-identical'. We now read $\exists X\varphi$ as "Either there are objects such that nothing is one of them or everything is self-identical." And this is indeed true, since the second disjunct ('everything is self-identical') is true.

So far we have merely hinted at Boolos' translation rules. Let us now consider them in detail. Let φ and ψ be formulas in which no first or second order constants and no second order descriptors occur. (We shall discuss them later.) Suppose, further, that each occurrence of any second order quantifier $\forall \delta$ has been replaced by an occurrence of $\neg \exists \delta \neg$. (This is merely a notational convenience.) Let α and β be first order variables and let δ be a second order variable. Let $f(\delta,\varphi)$ be the result of replacing each occurrence of $\delta\alpha$ (with δ *free*) in φ with an occurrence of $\alpha \neq \alpha$. (The occurrences of δ must be free because, otherwise, the interpretation function defined below will map theorems such as '$\exists z \exists X(\forall y \neg Xy \wedge \exists X\, Xz)$' into falsehoods.) We now introduce a translation function i that maps monadic second order formulas into assertions in a kind of quasi-English.

i($\neg \varphi$)	is	"it is not the case that i(φ)"
i($\varphi \to \psi$)	is	"i(φ) only if i(ψ)"
i($\forall \alpha \varphi$)	is	"every object$_\alpha$ is such that i(φ)"
i($\exists \delta \varphi$)	is	"either there are objects$_\delta$ such that i(φ) or i($f(\delta,\varphi)$)"
i($\alpha = \beta$)	is	"it$_\beta$ is identical to it$_\alpha$"
i($\delta \alpha$)	is	"it$_\alpha$ is one of them$_\delta$"

Attaching the subscripts α, β, and δ to the pronouns 'it' and 'them' allows us to keep track of the precise role those pronouns play in our quasi-English translation. In technical jargon: the subscripts allow us to handle complex cases of pronominal cross-reference. In less forbidding language: they allow us to determine which occurrences of 'it' and 'them' are meant to refer to the same things. Subscripting the word 'object' allows us to match up quantifiers with the pronouns which they bind.

Example: Let ψ be Theorem 1: '$\exists X \forall y\, Xy$'. Letting φ be '$\forall y\, Xy$', i(φ) is 'every object$_y$ is such that it$_y$ is one of them$_x$', and $f(X,\varphi)$ is '$\forall y\; y \neq y$', and i($f(X,\varphi)$) is 'every object$_y$ is such that it is not the case that it$_y$ is identical to it$_y$'. So i(ψ) is 'either there are objects$_x$ such that every object$_y$ is such that it$_y$ is one of them$_x$ or every object$_y$ is such that it is not the case that it$_y$ is identical to it$_y$' – which, with a little effort, can be seen to be equivalent to 'either there are objects such that everything is one of them or nothing is self-identical'. As desired, our translation of Theorem 1 is a true sentence of quasi-English (since it is a disjunction whose first disjunct is true).

Another example: Let ψ be Theorem 2: '$\exists X \forall y\, \neg Xy$'. Letting φ be '$\forall y\, \neg Xy$', i(φ) is 'every object$_y$ is such that it is not the case that it$_y$ is one of them$_x$', $f(X,\varphi)$ is '$\forall y\, \neg y \neq y$', and i($f(X,\varphi)$) is 'every object$_y$ is such that it is not the case that it is not the case that it$_y$ is identical to it$_y$'. So i(ψ) is 'either there are objects$_x$ such that every object$_y$ is such that it is not the case that it$_y$ is one of them$_x$ or every object$_y$ is such that it is not the case that it is not the case that it$_y$ is identical to it$_y$' – which, with a little more effort, can be seen to be equivalent to 'either there are objects such that nothing is one of them or everything is self-identical'. As desired, our translation of Theorem 2 is a true sentence of quasi-English (since, in this case, it is a disjunction whose second disjunct is true).

In fact *every* theorem of our system which contains no constants or descriptors will have a true (albeit grotesque) quasi-English translation. Since, with a little practice, we can learn to make sense of the quasi-English translations, it would be perverse to claim that the original second order sentences are unintelligible. If we decided to frame a set theory in a monadic second order language, we could not be justly accused of mystery-mongering. This point turns out to have immensely important consequences – which we shall explore in just a bit.

First, however, I must explain why Boolos' translation rules can be regarded as satisfactory even though they leave our constants and descriptors unaccounted for. When I introduced both our first and second order constants, I mentioned that I was doing so for reasons of technical convenience. My adoption of second order descriptors was similarly motivated. It happens that the rules for a second order deductive system of our sort can be formulated with relative ease in the presence of constants and descriptors. But, in principle, we can do without them. For our system is a conservative extension of monadic second order systems which lack constants and descriptors. Since we *can* do without constants and descriptors, it doesn't hurt for us to pretend that we *have* done without them. And this is essentially what we are doing in adopting Boolos' translation rules. Sentences featuring constants and descriptors furnish us with elegantly defined routes to theorems which lack them. Since it is those theorems which really interest us, we ignore the path we took to reach them.

4. Second Order Z

Let us now turn to the topic of second order set theories. It will be a great help to add all the remaining logical symbols from chapter 4 ('↔', '∧', '∨', '∃') to our second order language. (Don't worry: I won't burden you with new rules of inference governing these symbols.) If we are to formulate a set theory, we also need '∈'. We now let Z^2 (monadic second order Zermelo set theory) be the conjunction of the following sentences.

Pairing $\forall x \forall y \exists z \forall w (w \in z \leftrightarrow (w = x \lor w = y))$
Separation $\forall X \forall x \exists y \forall z (z \in y \leftrightarrow (z \in x \land Xz))$
Power Set $\forall x \exists y \forall z (z \in y \leftrightarrow \forall w (w \in z \rightarrow w \in x))$
Union $\forall x \exists y \forall z (z \in y \leftrightarrow \exists w (z \in w \land w \in x))$
Infinity $\exists x (\exists y (y \in x \land \neg \exists z \ z \in y) \land \forall y (y \in x \rightarrow \exists z (z \in x \land \forall w (w \in z \leftrightarrow w = y))))$
Extensionality $\forall x \forall y (\forall z (z \in x \leftrightarrow z \in y) \rightarrow x = y)$

I have omitted the empty set axiom here because it is derivable from the above version of Separation. I shall discuss the Axiom of Choice in a little while.

We say that a sentence φ is Z^2-*true* if the conditional ($Z^2 \rightarrow$ φ) comes out true when its logical components are interpreted via the Boolos translation into quasi-English and '∈' is assigned any meaning whatsoever. That is to say, the Z^2-truths are those sentences which are logical consequences of Z^2. (Logical consequence has nothing whatever to do with extra-logical content – hence the requirement that ($Z^2 \rightarrow$ φ) come out true no matter what content is attached to the non-logical symbol '∈'.) We say that a sentence φ is a Z^2-*theorem* if the conditional ($Z^2 \rightarrow$ φ) is a theorem of our second order deductive system. It follows from Gödel's First Incompleteness Theorem that there are Z^2-truths which are not Z^2-theorems. In fact, our deductive system is incorrigibly incomplete – for, no matter how we strengthen it, there will continue to be Z^2-truths which fail to be theorems (unless, of course, we render our system formally inconsistent). Second order logic is incomplete and incompletable in the following sense: deductive systems in which genuine proofs can be mechanically distinguished from bogus ones will never capture every form of inference which is warranted by the meaning of the second order logical symbols. Our deductive system has (I hope) served as a useful introduction to monadic second order logic. But it is not a fully adequate representation of monadic second order validity. If we let Boolos' translation rules supply us with an interpretation of our formal language, our grasp of monadic second order reasoning will ultimately depend on our mastery of plural quantification in English. This strikes me as quite solid ground on which to base the semantics of our logic.

Zermelo's Axiom of Choice (AC) and either the Continuum Hypothesis (CH) or its negation ($\neg$CH) are examples of Z^2-truths which are not Z^2-theorems. We shall discuss CH in our next section. Let us now consider the Axiom of Choice. AC states that every set s of non-empty, pairwise disjoint sets has a choice set — a choice set for s being a set whose membership features exactly one member of each member of s. In symbols:

$$\forall w((\forall x \in w \exists y \; y \in x \land \forall x,y \in w \forall z((z \in x \land z \in y) \to x=y)) \to$$
$$\exists z \forall x \in w \exists! y \in x \; y \in z).$$

Here $\forall \alpha \in \beta \varphi$ is an abbreviation of $\forall \alpha(\alpha \in \beta \to \varphi)$, $\forall \alpha,\beta \in \gamma \varphi$ is an abbreviation of $\forall \alpha \in \gamma \forall \beta \in \gamma \varphi$, $\exists! \alpha \varphi(\alpha)$ is an abbreviation of $\exists \alpha \forall \beta(\varphi(\beta) \to \beta = \alpha)$ (and is to be read as "there is a unique α such that $\varphi(\alpha)$"), and $\exists! \alpha \in \beta \varphi(\alpha)$ is an abbreviation of $\exists! \alpha(\alpha \in \beta \land \varphi(\alpha))$. Let's consider why this formula should be regarded as a Z^2-truth.

Let s be a set of non-empty, pairwise disjoint sets. Does it follow that there are objects among which exactly one member of each member of s occurs? That is, are there objects such that $\forall x \in s \exists! y \in x (y$ is one of those objects)? Note that I am not asking whether there is a *set* whose membership features such objects (i.e., whether $\exists z \forall x \in s \exists! y \in x \; y \in z$) — I am merely asking whether such objects exist. So, at the moment, I am stopping short of asking whether AC itself is true.

More generally, let $\mathcal{R}$ be a binary relation and c an object such that:

1. $\forall x(\mathcal{R}xc \to \exists y \; \mathcal{R}yx)$
2. $\forall x,y,z((\mathcal{R}xc \land \mathcal{R}yc \land \mathcal{R}zx \land \mathcal{R}zy) \to x=y)$.

Does it follow that there are objects such that $\forall x(\mathcal{R}xc \to \exists! y(\mathcal{R}yx \land (y$ is one of those objects)))? Note, again, that this is very different from the question of whether $\exists z \forall x(\mathcal{R}xc \to \exists! y(\mathcal{R}yx \land \mathcal{R}yz))$. It would be rash to answer this latter question without knowing quite a bit more about $\Re$, c, and the quantificational domain (i.e., the domain over which the quantifiers are taken to range). But what about the former question? *Are there* objects of the desired sort? The answer must be "yes" – as long as we wish our plural quantifier 'there are' to share the realist features of the classical singular quantifier 'there is'.

Let's examine why this is so. One is an antirealist with respect to a language L if one denies that every L-sentence is determinately either true or false and if, furthermore, one does so on the ground that our ability to *recognize* an L-sentence's truth is necessary for its *being* true.[87] On such a view, a singular existential generalization

There is an object which has the property F

is true just in case we can satisfy the conditions which we ourselves have laid down for its assertability. Typically, this will involve either specifying an object which can be shown to have the property F or indicating a procedure which, if carried out, would result in such a specification. A plural existential generalization

There are objects which have the property G

would, similarly, be true only if we either could specify objects which can be shown to have the property G or could indicate a procedure which would supply us with such a specification. A realist, on the other hand, would maintain that the truth-value of an existential generalization (whether singular or plural) is independent of our ability to *establish* its truth via specification of appropriate objects. On this view, there could indeed be objects which collectively have a certain property even if we are in principle unable to say just what objects they might be. Indeed, it is the essence of realism to distinguish sharply between the truth of an existential generalization and our ability actually to track down objects which witness to the truth of that generalization.

Returning now to our earlier question: Given premises 1 and 2, are there objects such that $\forall x(\Re xc \rightarrow \exists!y(\Re yx \wedge (y$ is one of those objects$)))$? The antirealist response would be "I don't know" – since we lack sufficient information to tell whether such objects could actually be specified. A realist, on the other hand, need not depend upon information of this sort. In fact, from a realist perspective, the only consideration which is directly germane to the truth of our proposition is precisely the truth of premises 1 and 2. For example, whether there are infinitely many objects which stand in $\Re$ to c is a matter of some importance to anyone who wishes actually to specify unique representatives of objects which stand in $\Re$ to objects which themselves stand in ℜ to c — but it is not a matter which a realist can admit as directly relevant to the *existence* of such representatives. The sort of thing which would, from a realist standpoint, rule out the existence of unique representatives is, for example, the existence of objects x, y, z which stand in ℜ to c and which have the following properties: distinct objects y' and z', and no others, stand in ℜ to x; at most the single object y' stands in ℜ to y; and at most the single object z' stands in ℜ to z. But this is precisely the sort of thing which is ruled out by premises 1 and 2. So, given 1 and 2, the realist should confidently assert that there are objects such that $\forall x(\Re xc \rightarrow \exists! y(\Re yx \wedge (y$ is one of those objects$)))$. For, from a realist perspective, the only objective conditions which could undermine the truth of this proposition are all ruled out by the truth of 1 and 2. This means that the decision to treat plural quantifiers in a realist manner commits one to an axiom (scheme) of choice for those quantifiers. Conversely, the adoption of such an axiom is an excellent (perhaps even the best available) way to express one's realist intentions within a language containing plural quantifiers. If, in light of George Boolos' work, we treat monadic second order logic as essentially a theory of plural quantification and if, furthermore, we wish to be thoroughgoing realists, then we should treat every formula of the form

$$\forall w((\forall x(\varphi xw \rightarrow \exists y \; \varphi yx) \wedge \forall x,y,z((\varphi xw \wedge \varphi yw \wedge \varphi zx \wedge \varphi zy) \rightarrow x=y)) \rightarrow \exists X \forall x(\varphi xw \rightarrow \exists! y(\varphi yx \wedge Xy)))$$

as a logical truth — in order to emphasize that, on our view, the existence of objects does not depend on our ability to construct a formula which characterizes them.

So far, we have only justified the adoption of a choice scheme for our second order quantifiers. We have yet to establish that the set theoretic AC is a Z^2-truth. As Boolos and Peter Simons have emphasized, plural variables are not any more prone to range over sets than are singular ones. Plural variables range plurally over singular objects; they do not range singularly over plural objects. So whether an instance of the scheme just cited above is set theoretic in character depends entirely on what formula is used to replace 'φ' and what quantificational domain is selected. It turns out that one such instance is of tremendous importance when introduced into appropriate set theories.

If we write '∈' in place of 'φ' in the above scheme, then we obtain:

$$\forall w((\forall x \in w \exists y \; y \in x \wedge \forall x,y \in w \forall z((z \in x \wedge z \in y) \rightarrow x=y)) \rightarrow \exists X \forall x \in w \exists! y \in x \; Xy).$$

One may initially fail to see why this formula is of particular set theoretic importance — for it is certainly not a standard set theoretic axiom of choice. Given a w which satisfies the antecedent, this formula does not by itself guarantee the existence of a choice set for w. Instead, it merely proposes that among objects which are already supposed to exist (i.e., the members of $\cup w$) there are some which collectively behave like the members of a choice set for w — and this is a relatively weak claim. Nonetheless, if this formula is accepted as a logical truth, it follows that AC is a logical consequence of Z^2. Briefly: given an appropriate w, we can extract a choice set from $\cup w$ by using second order Separation together with the above formula. Thus, the axiom of choice (in its usual, set theoretic guise) must be accepted by anyone who subscribes to Z^2 and, furthermore, regards the second order quantifiers of Z^2 as realist plural quantifiers.[88]

Note that a realist interpretation of plural set theoretic quantifiers is at least as defensible as a realist interpretation of *singular* ones. In any given language, plural variables range over the same objects as singular ones. If our latitude in referring singularly to objects via quantifiers is constricted only by the objective breadth of the quantificational domain, then our latitude in referring plurally need be no more constricted — that is, there are no grounds for additional constriction which should be compelling to anyone whose realism is otherwise unwavering. So a decision to interpret *only* singular quantifiers in a realist manner would be utterly unmotivated philosophically.

To complete my discussion of AC, I need only note that Boolos must intend the English plural quantifier 'there are' to be read in a realist way in the context of his translation rules — for, otherwise, his rules would not give a faithful rendering of the classical '$\exists X$'. Remember now that AC is Z^2-true if the conditional ($Z^2 \to$ AC) comes out true when its logical components are interpreted in Boolos' way and '$\in$' is assigned any meaning whatever. Given the realist character of Boolos' interpretation, the Z^2-truth of AC follows from the above discussion.

5. CH Is Second Order Decidable

In chapter 5, we saw that Paul Cohen and Abraham Robinson issued a challenge to their nonformalist colleagues: "Establish the respectability of your concept of mathematical set by articulating it in a way that allows the truth or falsity of CH to be decided." Since (for reasons to be discussed in our next chapter) either CH or $\neg$CH is Z^2-true, the Robinson-Cohen challenge would seem to have been decisively answered: the required articulation is Z^2. Granted, we do not at the moment know whether it is CH or ¬CH which is Z^2-true. But we do seem to have expressed a notion of set which is sufficiently determinate to fix the truth or falsity of CH. How might a skeptical formalist respond to this?

Our use of Boolos' translation rules has forestalled one popular formalist response. Monadic second order quantifiers are usually taken to range over *sets*. More precisely: in the standard semantics, first order quantifiers are taken to range over the members of some set $\mathfrak{D}$ while monadic second order quantifiers are taken to range over the subsets of $\mathfrak{D}$. On this view, $\exists X\, \varphi(X)$ means that there is a subset of $\mathfrak{D}$ which satisfies the predicate φ.[89] A formalist who considers the notion of mathematical set to be mysterious would, naturally, feel less than enlightened by this account. From a formalist point of view, it's bad enough that the second order logical consequence relation cannot be captured in a formal deductive system; an attempt to legitimize this alleged "logic" by appealing to the repugnant notion of set only makes matters worse. Against this background, we can sympathize with formalists who regard second order set theories as miserable displays of mystery-mongering and question-begging.

But look how the scenario changes when we explicate monadic second order logic via Boolos' translation rules: it becomes much more difficult for a formalist to argue for the unintelligibility of second order quantifiers. Naive notions of set are relatively easy targets. But how is one to establish that the ordinary English plural quantifier 'there are' (a locution we all constantly employ) is fundamentally incomprehensible? The answer is obvious: one is never going to establish any such thing. This line of formalist attack has been effectively blocked.

Yet we still must face serious problems arising from Gödel's First Incompleteness Theorem. Recall that there are Z^2-truths which are not Z^2-theorems. Indeed, our supply of truths will surpass our supply of theorems no matter how we augment our deductive system. Consider what this means in connection with Cantor's Continuum Hypothesis. Since neither CH nor ¬CH are Z^2-theorems, we should assert that CH or ¬CH is Z^2-true only if we are prepared to argue that our notion of set theoretic truth is richer than our notion of set theoretic proof. Indeed, we must be prepared to argue that our notion of set theoretic truth is richer than *any* notion of set theoretic proof we could possibly possess. For even if we strengthened our set theoretic deductive system in such a way as to make CH or ¬CH into a theorem, there would still be set theoretic truths which are not theorems.

Here we seem to run afoul of the Dummettian argument in chapter 5. According to Claim 7 in that argument, a set theoretic proposition's *truth* conditions can be no better defined than the distinction between adequate and inadequate *proofs* of that proposition. Yet aren't we now claiming to have fixed truth conditions for CH without at all clarifying the conditions under which CH could be considered proven or disproven (since our alleged recognition that either CH or $\neg$ CH is Z^2-true seems not to have brought us any closer to a proof or disproof of CH)? Aren't we claiming to have done something which is, in fact, impossible?

Perhaps not. We do *not* claim to understand thoroughly the conditions under which CH is true. We claim merely to possess a theory describing a world in which either CH or $\neg$ CH is true. If we discovered that it is, say, CH which is true in this world, we would *then* have a good idea of the conditions which render CH true and would be in a position to formulate a deductive system in which CH is provable. To put the matter somewhat differently: we claim to possess a picture whose content is entirely definite, but not entirely explicit. Part of this picture's content appears by implication only (like the sun in a Turner or Monet study of light). Either CH or $\neg$ CH figures in this implicit content; and we claim to have (in principle) enough explicit information to figure out which of the two it is. Our picture thus poses a puzzle which has only one correct solution, a solution which would simultaneously enrich our conception of set theoretic truth and set theoretic proof. It is in this sense that Z^2 is said to fix the truth value of CH.

We are not in the clear yet, however. For if we take seriously the compelling view of mathematics at the core of the Dummettian argument, we may be forced to admit that the truth value of CH could not have been fixed even in the relatively weak sense described above. We insisted, in chapter 5, that mathematics is a public, communal enterprise. A grasp of mathematical meaning must, accordingly, always be publicly exhibitable – either by explicit definition or by correct use. Since proofs are the primary arena for the employment of set theoretic expressions, we decided that set theoretic meanings and, hence, set theoretic truth conditions are implicitly displayed in overt set theoretic proof practices. And we concluded that a proposition such as CH, whose truth conditions are left significantly undetermined by current proof practices, is not fit to bear a determinate truth value. Any claim to have "fixed" such a value is, therefore, unfounded.

If we are to sidestep this powerful argument we must somehow loosen the link between mathematical truth and mathematical proof; we must allow for the determination of a set theoretic proposition's truth value by something other than set theoretic proof practices. The following view of mathematical meaning *might*, if successfully developed, do the job. (1) If φ is a non-logical mathematical symbol other than '∈', then our knowledge of φ's meaning consists in our ability to define φ set theoretically. (2) Our knowledge of what '∈' means consists, first, in our ability to recite the axioms of a second order set theory and, second, in our grasp of the logical vocabulary of that theory. (3) Our grasp of the theory's logical vocabulary consists, first, in our ability to apply Boolos' translation rules (or others like them) to monadic second order formulas and, second, in our grasp of the logical vocabulary occurring in the English translations which are thus generated.

Thesis 1 is a form of set theoretic reductionism – a position which we touched upon in chapter 1 (and elsewhere). Thesis 2 is a form of set theoretic structuralism – a position we shall consider in our next two chapters. It is the third thesis which most interests us at the moment. It is intended to break the proof/truth link by rooting our knowledge of mathematical meaning outside the domain of mathematical practice. Theses 1 and 2 imply that our grasp of mathematical expressions flows from our grasp of the symbolism of monadic second order logic. Thesis 3 then traces our grasp of the logical symbolism back to our mastery of ordinary English locutions. The upshot is meant to be that the truth values of mathematical sentences are not determined by overt mathematical proof practices, but rather by patterns of ordinary usage which allow locutions such as plural quantifiers to be given a realist reading. This would mean that our commonsense mastery of plural reference is crucial to our understanding of mathematical set theory. This would *not* imply that such mastery is in itself sufficient to supply anyone with an adequate notion of mathematical set. (So the main conclusion of our chapter 3 would still stand.)

I have offered only the barest sketch of an antiformalist program which no one (to my knowledge) has worked out fully. My point in doing so is merely to suggest that a suspension of belief in the formalist conclusions of chapter 5 may not be hopelessly irrational. If formalism were inescapable, the next two chapters would be largely pointless. So, lest my readers set this volume down, I have tried to offer the opponents of formalism a glimmer of hope.

We shall now begin to consider the approach to Cantorian set theory which is in the best position to compete seriously with formalism. I mean the view that the primary objects of set theoretic research are pure structures. In chapter 8, we review the sort of axiomatizations which are particularly appealing from a structuralist point of view. In chapter 9, we examine some philosophical articulations and defenses of mathematical structuralism.

VIII

ITERATIVE HIERARCHIES

1. Set Formation

In chapter 5, we explored the view that *theories* are the only legitimate objects of set theoretic research. We shall now see how set theorists can be viewed as scientists who investigate the properties of *structures* known as "iterative hierarchies." Very roughly, an iterative hierarchy is an array of sets regarded as the product of an iterated process of set formation which begins with the empty set and then proceeds through a well-ordered series of stages, each stage containing sets whose members appear at prior stages. There are all sorts of theories about iterative hierarchies, each of which can be thought to characterize *the* "real, honest-to-goodness, genuine iterative hierarchy." But, rather than endorsing a particular view about "the" iterative hierarchy, we shall speak of iterative hierarch*ies* (in the plural). We shall view set theorists as devotees of a family of distinct structures, each of which will be accorded the title of *an* iterative hierarchy.

It is common to talk about iterative hierarchies coming into existence through a *human, mental* process of set formation *stretched out over time*. Since the hierarchies under consideration turn out to have enormous and enormously complex infinitary structures, this talk could be taken literally only if one gave a very peculiar and implausible account both of the nature of time and of the extent of human mental powers.[90] I acknowledge that the image of temporal, mental set formation may be a useful *picture*. But it is one which I feel little inclination to take seriously.

This implausible picture is tempting because, in an iterative hierarchy, sets are ordered in such a way that the members of a set are always prior in the ordering to the set of which they are the members. So if one takes this priority to be temporal, one acquires the natural image of a set being produced only after all of its members have been produced. And this allows one to make a certain amount of sense of the particular ways sets are arranged in an iterative hierarchy. For example, the position of the empty set at the bottom (or at the top – depending on how you look at it) of every iterative hierarchy becomes fairly natural: after all, the empty set is the one set which need not be posterior to its members – for the simple reason that it has none. The very expression 'iterative hierarchy' harks back to the notion that the universe of set theory is created by first forming the empty set and then iterating the process of set formation. More precisely: At time 0, one forms all the sets whose members are sets which have already been formed; that is, since no sets have been formed, one forms the empty set:

$$\emptyset.$$

At time 1, one again forms all the sets whose members are sets which have already been formed; that is, one again forms the empty set and, further, since the empty set has already been formed, one forms the set whose only member is the empty set:

$$\emptyset, \{\emptyset\}.$$

At time 2, one yet again forms all the sets whose members are sets which have already been formed:

$$\emptyset, \{\emptyset\}, \{\{\emptyset\}\}, \{\emptyset,\{\emptyset\}\}.$$

And so on. Thus, at time 0, one forms 2^0 sets – that is, one forms just one set. At time 1, one forms 2^1 (=2) sets; at time 2, 2^2 (=4) sets; at time 3, 2^4 (=16) sets; at time 4, 2^{16} (=65,536) sets. And, at time 5, one forms $2^{65,536}$ (=???!) sets.

Of course, one really does nothing of the sort. Even if one managed to make sense of the notion of our "forming sets," one would be hard pressed indeed to explain how it is even *possible* for anyone to form $2^{65,536}$ sets. It would be harder still to establish that anyone *actually* does this. And note: we have so far dealt only with what is supposed to happen at very small finite time stages. One would also have to account for the formation of truly monstrous numbers of sets at time stages much later than stage 5. And I do mean *much* later. Depending on what iterative hierarchy one is committed to, one might even have to account for the formation of sets at a "time" stage α – where α is a remote, complex, *infinite* ordinal number. So, as I already mentioned, this picture of temporal set formation can trap one in a highly questionable view not only of human mental powers, but of the structure of time as well. I shall take it for granted from now on that this picture is merely a metaphor.

2. Well-Ordered Structures

Iterative hierarchies are structures determined by the relation $\in$. To different theories about $\in$ there correspond different iterative hierarchies. But, in every case, $\in$ breaks the universe up into a well-ordered sequence of stages between the members of which complex and mathematically interesting relations hold. A well-ordering is a transitive, irreflexive, connected relation R such that every collection of objects contains a member to which no other member stands in R. That is, if x,y,z are among the objects well-ordered by R, then:

$$\neg Rxx$$
(Irreflexivity)
$$(Rxy \land Ryz) \to Rxz$$
(Transitivity)
$$Rxy \lor Ryx \lor x=y$$
(Connectedness).

Further, given any objects at all, one of them will be "R-least." Suppose, for example, that the predicate 'P' is satisfied by at least one object. (That is, suppose that $\exists x\, Px$.) Then, if R is a well-ordering, there is an "R-least" object which satisfies 'P':

$$\exists x(Px \land \neg \exists y(Py \land Ryx)).$$

An example may make the expression 'R-least' clearer. If we let R be the "less than" relation over the natural numbers and let 'Px' mean that x is prime, then the formula

$$\exists x(Px \land \neg \exists y(Py \land Ryx))$$

would say that there is a prime number such that no prime number is less than it – that is, there is a least prime number. With this sort of example in mind, we speak of "R-least" objects where R is any well-ordering (not necessarily a standard "less than" relation over a system of numbers). To be the R-least member of a collection is to have no other member of that collection stand in R to oneself.

To return to our discussion of iterative hierarchies: $\in$ is not itself a well-ordering of the universe, but it can be used to divide up the universe into certain collections of objects, to define a well-ordering of these collections, and to define certain interesting relations between the members of these collections. Stage 0 of the well-ordering contains a single object $\emptyset$ to which nothing stands in $\in$. Stage 1 contains the object $\{\emptyset\}$ to which $\emptyset$ stands in $\in$ (that is, $\emptyset \in \{\emptyset\}$). Stage 2 contains the object $\{\{\emptyset\}\}$ to which $\{\emptyset\}$ stands in $\in$ and the object $\{\emptyset,\{\emptyset\}\}$ to which both $\emptyset$ and $\{\emptyset\}$ stand in $\in$. This yields the picture:

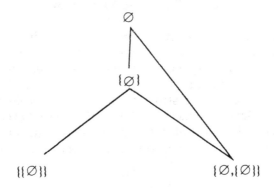

where the (more or less) vertical lines represent the ∈ relation. This differs from the account given above merely in that objects from prior stages do not reappear at later ones. In either case, we can say that an object x is introduced prior to an object y (that is, the stage at which x first appears is prior in the well-ordering to the stage at which y first appears) whenever $x \in y$.

An iterative hierarchy is mathematically interesting because it contains mathematically interesting sub-structures. For example, an iterative hierarchy with infinitely many stages will contain the sub-structure

$$\begin{array}{c} \varnothing \\ \{\varnothing\} \\ \{\{\varnothing\}\} \\ \{\{\{\varnothing\}\}\} \\ \{\{\{\{\varnothing\}\}\}\} \\ \cdots \end{array}$$

which we can recognize as essentially Ernst Zermelo's model of the natural numbers. (Cf. the interpretation of the theory of the successor function given in chapter 6.) A sufficiently large hierarchy will also contain the sub-structure

$$\begin{array}{c} \varnothing \\ \{\varnothing\} \\ \{\varnothing,\{\varnothing\}\} \\ \{\varnothing,\{\varnothing\},\{\varnothing,\{\varnothing\}\}\} \\ \{\varnothing,\{\varnothing\},\{\varnothing,\{\varnothing\}\},\{\varnothing,\{\varnothing\},\{\varnothing,\{\varnothing\}\}\}\} \\ \cdots \\ \{\varnothing,\{\varnothing\},\{\varnothing,\{\varnothing\}\},\ldots\} \\ \{\varnothing,\{\varnothing\},\{\varnothing,\{\varnothing\}\},\ldots,\{\varnothing,\{\varnothing\},\{\varnothing,\{\varnothing\}\},\ldots\}\} \\ \cdots \end{array}$$

This is essentially Dimitry Mirimanoff's model of the finite and infinite ordinal numbers (a model which is generally attributed to von Neumann).[91] The basic idea is that each ordinal is the set of all prior ordinals (that is, each ordinal is such that all and only the prior ordinals stand in $\in$ to it). So the 0th ordinal is the empty set. And the first infinite ordinal (usually called "ω") is the set of all finite ordinals. The ordinal numbers are supposed to be abstract representatives of the various types of well-orderings. (Whereas the cardinal numbers are meant to be abstract representatives of *magnitudes*.) Thus it is customary to say that ω is the *order type* of, for example, the natural numbers. For the natural numbers, when ordered by $<$, are isomorphic to (that is, have the same structure as) the members of ω, when ordered by $\in$. And ω is taken to be the representative or "order type" of all the structures isomorphic to it. It is important to remember at this point that an iterative hierarchy is itself a well-ordered sequence of stages. This raises the important question of whether the structure of the whole sequence of stages is reflected in an order type contained within one such stage and, indeed, whether a well-ordered sub-structure within the hierarchy of stages might not be *more* complex than the whole sequence of stages. More on this later.

3. Choosing a Theory

It is now almost time to present some theories of iterative hierarchies. This requires that we consider two questions. First, what is to be the overall structure of our hierarchy of stages; that is, what order type should the whole sequence of stages embody? Second, what contents should the stages have; that is, how should we make out the process of "set formation" which supplies each stage with its contents? Let us tackle the second question first. Each stage of an iterative hierarchy is supposed to contain all the sets which do not themselves appear at prior stages, but whose members are sets which appear at prior stages. In more neutral language, each stage is supposed to contain "all possible new combinations" of objects from previous stages. So given any stage α and any combination K of objects from prior stages, there is an object x (at either α or a prior stage) such that all and only the objects in K stand in $\in$ to x. What account are we to give of this notion of a "combination"?

Chapter 7 has put us within reach of a most satisfactory answer to this question. For the notion of "combination" at work here can be well expressed within monadic second order logic. And the intelligibility and philosophical respectability of this logic has been guaranteed by George Boolos' translation rules. To review: By exploiting the familiar devices of plural reference and plural quantification, Boolos demonstrates that monadic second order variables can be taken to range over exactly the same objects as first order variables without any loss of deductive power. The two sorts of variables are just taken to do their ranging in different ways. Suppose, for example, that we intend our first order variables to refer to sets within a hierarchical array. Then we can use second order variables to capture the notion of "possible combinations" while, simultaneously, letting these variables refer to the same hierarchically ordered sets as the first order ones.

We daily refer to objects either one at a time ("Tom lifted a girder") or many at a time ("Tom, Dick, and Harry lifted a girder"). What interests us at the moment is that the plural form of reference need not be interpreted as a way of referring to special plural objects (such as sets). We can instead regard it as a special way of referring to individuals (more than one at a time). Similar remarks apply to the plural quantifier 'There are'. Instead of taking it to range over plural objects of some sort, we can take it to range plurally over the same domain of individuals as the first order singular quantifier 'There is'. So if we interpret monadic second order logic using plural quantification, we can establish that this logic commits us to no special plural objects. This is what Boolos does.

In order to remind ourselves how this works, consider again the assertion: "Given any stage α and any combination K of objects from prior stages, there is an object x such that all and only the objects in K stand in $\in$ to x." In our formal theory, we shall express this by using the second order variable 'X' and the binary relation symbol 'F' (where '$Fy\beta$' means that y appears at stage β):

$$\forall X \forall \alpha \exists x \forall y (y \in x \leftrightarrow (Xy \land \exists \beta < \alpha \, Fy\beta)).$$

('$\exists \beta < \alpha Fy\beta$' means that $\exists \beta(\beta < \alpha \land Fy\beta)$.) To regard '$\forall X$' as a plural quantifier is, roughly, to interpret the original English assertion as

> Given any stage α and any objects from prior stages, there is an x to which all and only those objects stand in $\in$.

In order to be entirely precise, we must really give the somewhat more unwieldy interpretation:

> Given any stage α and any objects from prior stages, there is an x to which all and only those objects stand in $\in$ and there is also an x to which nothing stands in $\in$.

The additional clause is required because the second order variable 'X' might fail to refer to anything at all or, to put it differently, the collection of objects to which the variable refers might be empty. (As we saw, it is a theorem of second order logic that $\exists X \forall x \, \neg Xx$.) But a plural quantifier is always thought to refer to objects. (The claim that there are coins in my pocket would be false if my pocket were actually empty.) So if we use a plural quantifier to interpret a second order one, we must take special steps to capture the whole meaning of the second order version.

We have been dealing with the issue of the contents of each stage in our iterative hierarchy. Since a stage is supposed to contain "all possible new combinations" of objects from previous stages, we have tried to explicate the notion of a "possible combination." Monadic second order logic, construed as a theory of plural quantification, captures this notion quite well. Now that we have an adequate grasp of the interior of each stage (or, at least, know what tools we are going to use to describe the interior of each stage), we need to consider the structure of the whole sequence of stages.

Since an iterative hierarchy is a *well-ordered* sequence of stages, an account of the overall structure of such a hierarchy can take the form of a theory of ordinals (that is, a theory of abstract well-orderings). Accordingly, our own accounts of iterative hierarchies will begin with descriptions of ordinals – ordinals which are to be correlated (or even identified) with the stages of the hierarchy. I am going to present two theories of iterative hierarchies, the first of which is essentially a version of second order Zermelo Set Theory. Serious deficiencies in this first theory will lead us to formulate a version of second order Zermelo-Fraenkel Set Theory.

4. Theory T_1

'X','Y','Z','X'','Y'','Z'', ... will be our monadic second order variables. We shall adopt a two-sorted system of first order variables. 'α','β','γ', 'α'','β'','γ'', ... are to range over our ordinals. 'x','y','z','x'','y'','z'', ... are to range over the objects appearing within the stages of the iterative hierarchy. We shall assume that if s is a first order variable of either sort, φ is a predicate in which 'X' does not occur free, and $t_1, \ldots, t_n$ are the free variables of φ, then

$$\forall t_1 \ldots \forall t_n \exists X \forall s (Xs \leftrightarrow \varphi)$$

is a theorem of our logic. (That is, given any predicate, it is possible, via a second order variable, to refer plurally to the objects which sat-

122 ITERATIVE HIERARCHIES

isfy that predicate.) This will allow us to interchange predicates and second order variables in the manner discussed in chapter 7.

Here are some axioms about our ordinals:

> *Axiom 1.1:* The ordinals are well-ordered by $<$. That is:
> 1.11 (Irreflexivity) $\forall \alpha \; \neg \; \alpha < \alpha$
> 1.12 (Transitivity) $\forall \alpha \forall \beta \forall \gamma ((\alpha < \beta \wedge \beta < \gamma) \rightarrow \alpha < \gamma)$
> 1.13 (Connectedness) $\forall \alpha \forall \beta (\alpha < \beta \vee \beta < \alpha \vee \alpha = \beta)$
> 1.14 $\forall X (\exists \alpha X \alpha \rightarrow \exists \alpha (X \alpha \wedge \neg \exists \beta (X \beta \wedge \beta < \alpha)))$

Note that we have already been able to put a second order quantifier to good use in Axiom 1.14. This allows Axiom 1.1 to characterize the basic order properties of $<$ in a truly satisfactory way. Axiom 1.14 says that every collection of ordinals contains a least member. More precisely, there are no ordinals among which there is no least.

> *Axiom 1.2:* There is a limit ordinal; i.e., $\exists \alpha (\exists \beta \; \beta < \alpha \wedge \forall \beta < \alpha \exists \gamma < \alpha \; \beta < \gamma)$.

A limit ordinal has a predecessor, but has no *immediate* predecessor. That is, if α is a limit ordinal, then, first, some ordinal β is prior to α and, second, corresponding to every ordinal β which is prior to α, there is an ordinal γ which is both prior to α and posterior to β. This second clause means that another predecessor always slips between a limit ordinal and one of its predecessors. So a limit ordinal will have *infinitely many* predecessors. For if an ordinal has only finitely many predecessors, it will be *immediately* preceded by the greatest of these (and if there is no greatest, it must actually have infinitely many predecessors). Although every limit ordinal is infinite (that is, has infinitely many predecessors), it does not follow that every infinite ordinal is a limit ordinal. For if a limit ordinal α had an immediate successor, then that successor would be infinite (because it is preceded by the infinitely many predecessors of α), but it would not be a limit ordinal (since α is its immediate predecessor).

We already know that each ordinal prior to the first limit ordinal has a successor. But we need a new axiom to guarantee that the first limit ordinal itself has a successor:

> *Axiom 1.3:* Every ordinal has a successor; i.e., $\forall \alpha \exists \beta \; \beta > \alpha$.

Axioms 1.2 and 1.3 are meant to guarantee that the sequence of ordinals is not too small. They imply that our hierarchy of stages will have at least the structure

$$0, 1, 2, 3, \ldots \omega, \omega+1, \omega+2, \omega+3, \ldots$$

THEORY T₁ 123

Axioms 1.1 and 1.3 together imply that every ordinal has an *immediate* successor:

Theorem 1: $\forall\alpha\exists\beta>\alpha\forall\gamma>\alpha\ \gamma\geq\beta$.

Proof: By Axiom 1.3, $\exists\beta\ \alpha<\beta$. By Axiom 1.14, pick the first such β. Then, by Axiom 1.13, $\forall\gamma>\alpha\ \gamma\geq\beta$; that is, every ordinal posterior to α is either posterior or identical to β. (In other words, no ordinal slips in between α and β.)

We now state some axioms about the "formation" of sets. '$Fx\alpha$' is to mean that x is "formed" at stage α — or, less metaphorically, that x belongs to or appears at stage α.

Axiom 2.1: $\forall x\forall\alpha(Fx\alpha \leftrightarrow (\forall y\in x\exists\beta<\alpha\ Fy\beta \land \forall\beta<\alpha\exists y\in x\forall\gamma<\beta\ \neg Fy\gamma))$.

Axiom 2.1 says that each set is formed as soon as all the sets belonging to it have been formed and not before. The clause '$\forall y\in x\exists\beta<\alpha\ Fy\beta$' says that every member of x appears at a stage prior to α. The clause '$\forall\beta<\alpha\exists y\in x\forall\gamma<\beta\ \neg Fy\gamma$', which is equivalent to '$\neg\exists\beta<\alpha\forall y\in x\exists\gamma<\beta\ Fy\gamma$', says that there is no stage β prior to α such that all the members of x appear at stages prior to β. So together the two clauses say that α is the first stage prior to which all the members of x appear. And the whole axiom says that a set x appears at a stage α if and only if α is the first stage prior to which all the members of x appear. This implies that no set can be formed at more than one stage:

Theorem 2: $\forall x\forall\alpha\forall\beta((Fx\alpha \land Fx\beta)\rightarrow\alpha=\beta)$.

Proof: Suppose that $Fx\alpha$, $Fx\beta$, and $\beta<\alpha$. Then, by Axiom 2.1 (left to right), $\exists y\in x\forall\gamma<\beta\ \neg Fy\gamma$. But this leads to a contradiction. For if $y\in x$, then, again by Axiom 2.1 (left to right), $\exists\gamma<\beta\ Fy\gamma$. We get a similar result if we assume that $Fx\alpha$, $Fx\beta$, and $\alpha<\beta$. We may conclude, by Axiom 1.13, that $\alpha=\beta$ whenever $Fx\alpha$ and $Fx\beta$. In less technical language: If α and β are each the first ordinal prior to which all the members of x appear, then α must be identical to β. For, otherwise, one would be prior to the other. And then only the prior one would be the *first* ordinal with the property under discussion.

Using a second order quantifier, we can express the notion that sets of sets are formed through some sort of selection of sets from an initial segment of the iterative hierarchy:

Axiom 2.2: $\forall x\exists X\exists\alpha\forall y(y\in x \leftrightarrow (Xy \land \exists\beta<\alpha\ Fy\beta))$.

This axiom says that, given any set x, either x is empty or there are sets appearing at stages prior to some stage α such that all and only those sets are members of x. This allows us to show that every set appears at some stage:

Theorem 3: $\forall x \exists \alpha\ Fx\alpha$.

Proof: By Axioms 1.14 and 2.2, let α be the first ordinal such that $\forall y(y \in x \leftrightarrow (Xy \wedge \exists \beta < \alpha\ Fy\beta))$. We want to apply Axiom 2.1 to establish that $Fx\alpha$. Clearly $\forall y \in x \exists \beta < \alpha\ Fy\beta$. Suppose that $\beta < \alpha$ and that $\forall y \in x \exists \gamma < \beta\ Fy\gamma$. Then clearly $\forall y(y \in x \to (Xy \wedge \exists \gamma < \beta\ Fy\gamma))$. On the other hand, if $(Xy \wedge \exists \gamma < \beta\ Fy\gamma)$, then, by Axiom 1.12, $(Xy \wedge \exists \gamma < \alpha\ Fy\gamma)$ and, hence, $y \in x$. So $\forall y(y \in x \leftrightarrow (Xy \wedge \exists \gamma < \beta\ Fy\gamma))$ – contradicting the minimality of α. So $\forall \beta < \alpha \exists y \in x \forall \gamma < \beta\ \neg Fy\gamma$. Thus, by Axiom 2.1 (right to left), $Fx\alpha$. In less technical language: We let α be the first stage prior to which there appear sets such that they and only they are members of x. It follows that α is the first stage prior to which all the members of x appear. So x appears at stage α.

By using a second order quantifier, we can capture the notion that at every stage each possible new combination of previously formed sets is gathered into a set.

Axiom 2.3: $\forall X \forall \alpha \exists x \forall y(y \in x \leftrightarrow (Xy \wedge \exists \beta < \alpha\ Fy\beta))$.

This axiom says that, given any sets appearing prior to some stage α, those and only those sets are members of some set x (that is, the existence of a set whose members are those and only those sets is guaranteed). This allows us to prove that at every stage at least one set appears:

Theorem 4: $\forall \alpha \exists x\ Fx\alpha$.

Proof: Suppose that $\forall y(Xy \leftrightarrow \exists \beta < \alpha\ Fy\beta)$ and, by Axiom 2.3, that $\forall y(y \in x \leftrightarrow (Xy \wedge \exists \beta < \alpha\ Fy\beta))$ – i.e., $\forall y(y \in x \leftrightarrow \exists \beta < \alpha\ Fy\beta)$. We want to show that x cannot be formed either before or after α. Suppose that $\gamma > \alpha$ and that $Fx\gamma$. Then, by Axiom 2.1, $\exists y \in x \forall \beta < \alpha\ \neg Fy\beta$ – which is absurd given our choice of x. On the other hand, suppose that $\gamma < \alpha$ and that $Fx\gamma$. Then $\exists \beta < \alpha\ Fx\beta$ and, thus, $x \in x$. But, by Axiom 2.1, $\forall y \in x \exists \beta < \gamma\ Fy\beta$. So $\exists \beta < \gamma\ Fx\beta$ and hence, by Axiom 1.1, $\exists \beta(\beta \neq \gamma \wedge Fx\beta)$ – contradicting Theorem 2. We conclude, by Axiom 1.13 and Theorem 3, that $Fx\alpha$. In less technical language: We let x be the set of all sets which appear at stages prior to α. It follows that α is the first stage prior to which all the members of x appear. (For if x appeared at a stage later than α, then some member of x would not appear prior to α – contrary to our assumption about the contents of x. And if x appeared at a stage earlier than α, then x would be a member of itself and, thus, would have to ap-

THEORY T_1 125

pear at a stage prior to a stage at which it appears — contradicting Theorem 2.) So x appears at stage α.

We shall adopt a standard version of the extensionality axiom.

Axiom 2.4: $\forall x \forall y (\forall z(z \in x \leftrightarrow z \in y) \rightarrow x=y)$.

We are now in a position to show that Zermelo's axioms are theorems of our system.

Theorem 5 (Pairing): $\forall x \forall y \exists z \forall w(w \in z \leftrightarrow (w=x \lor w=y))$.

Proof: By Axiom 1.13 and Theorem 3, suppose that x and y appear no later than stage α — that is, $\forall w((w=x \lor w=y) \rightarrow \exists \gamma \leq \alpha \ Fw\gamma)$. By Axiom 1.3, suppose that $\alpha < \beta$. Finally, by Axiom 2.3, suppose that $\forall w(w \in z \leftrightarrow ((w=x \lor w=y) \land \exists \gamma < \beta \ Fw\gamma))$. Then, by Axiom 1.12, $\forall w(w \in z \leftrightarrow (w=x \lor w=y))$. In less technical language: We let z be the set of all sets which appear prior to β and which are identical either to x or to y. Then only x and y can be members of z. But does z indeed have any members? Yes: both x and y are members of z since they both appear no later than α and, hence, appear prior to β.

The idea here is that the pair set $\{x,y\}$ will appear at the first stage prior to which both x and y have appeared. In general, in order to prove that a certain set exists, we need only show that there is a stage prior to which all the members of that set have appeared.

Theorem 6 (Union): $\forall x \exists y \forall z(z \in y \leftrightarrow \exists w(z \in w \land w \in x))$.

Proof: By Theorem 3 suppose that $Fx\alpha$. Then, by Axiom 2.1, if $w \in x$, then $\exists \beta < \alpha \ Fw\beta$. So, by Axioms 1.12 and 2.1, if $\exists w(z \in w \land w \in x)$, then $\exists \beta < \alpha \ Fz\beta$. By Axiom 2.3, suppose that $\forall z(z \in y \leftrightarrow (\exists w(z \in w \land w \in x) \land \exists \beta < \alpha \ Fz\beta))$. Then $\forall z(z \in y \leftrightarrow \exists w(z \in w \land w \in x))$. In less technical language: If the members of x appear prior to stage α, then so do the members of members of x. So if z is the set of all members of members of x, there is a stage (namely, α) prior to which all the members of z appear — which, as I have just noted, is what we want to show.

Theorem 6 guarantees the existence of $\cup x$ — the set of all members of members of x. Theorems 5 and 6 together guarantee the existence of the sets resulting from the usual Boolean operation of union. For $\cup\{x,y\}$ is the set of all sets which are members either of x or of y — that is, $\cup\{x,y\}$ is the union of x and y (usually designated by '$(x \cup y)$').

Theorem 7 (Empty Set): $\exists x \forall y \ y \notin x$.

Proof: By Axiom 1.14, let α be the very first ordinal. And, by Axiom 2.3, suppose that $\forall y(y \in x \leftrightarrow \exists \beta < \alpha \ Fy\beta)$. Then $\forall y \ y \notin x$. In less technical language: The sets of all sets which appear prior to the very first stage is empty (since no sets appear prior to the very first stage).

If we let '∅' designate the empty set, Theorems 5, 6, and 7 guarantee the existence of all the members of the infinite sequence generated from the empty set by repeatedly applying the operation $\mathcal{F}(x) = (x \cup \{x\})$; that is,

$$\emptyset, \emptyset \cup \{\emptyset\}, \emptyset \cup \{\emptyset\} \cup \{\emptyset \cup \{\emptyset\}\}, \ldots$$

or, in different notation,

$$\emptyset, \{\emptyset\}, \{\emptyset, \{\emptyset\}\}, \ldots$$

The reader may recall that this is Mirimanoff's model of the finite ordinals (each ordinal being the set of all previous ordinals). So we have established that all the finite Mirimanoff ordinals exist. But we have yet to show that there is a set to which all of them belong.

Theorem 8 (Infinity): $\exists x(\emptyset \in x \wedge \forall y(y \in x \to \mathcal{F}(y) \in x))$.

Proof: By Axiom 1.2, let α be a limit ordinal. And, by Axiom 2.3, suppose that $\forall y(y \in x \leftrightarrow \exists \beta < \alpha \ F y \beta)$ — that is, suppose that x is the set of all sets which appear prior to α. By Axiom 2.1, ∅ appears at the very first stage. So ∅ appears prior to α and, thus, is a member of x. Now suppose $y \in x$. Then y appears prior to α. We need to show that $\mathcal{F}(y)$ also appears prior to α. Suppose that y appears at stage β. $\mathcal{F}(y)$ consists of the members of y (which, by Axiom 2.1, all appear prior to β) and y itself (which appears at β). So, by Axiom 2.1, $\mathcal{F}(y)$ appears at the first stage posterior to β (since that would be the first stage prior to which all the members of $\mathcal{F}(y)$ appear). $\mathcal{F}(y)$ cannot, then, appear at stage α (since α is a limit ordinal and, so, cannot have an immediate predecessor). Hence $\mathcal{F}(y)$ appears prior to α and, thus, is a member of x. As desired, we have shown that ∅ is a member of x and that x is closed under the operation $\mathcal{F}$.

In the Mirimanoff model, ω (the first infinite ordinal) is the set of all finite ordinals. We have just seen that there is a set which includes all the finite ordinals. But is there a set whose members are exactly the finite ordinals? The next two theorems will help us prove that there is.

Theorem 9 (Separation): $\forall X \forall x \exists y \forall z(z \in y \leftrightarrow (z \in x \wedge Xz))$.

Proof: By Theorem 3, suppose that $Fx\alpha$. By Axiom 2.3, $\exists y \forall z(z \in y \leftrightarrow (z \in x \wedge Xz \wedge \exists \beta < \alpha \ Fz\beta))$. By Axiom 2.1, $\forall z \in x \exists \beta < \alpha \ Fz\beta$. So $\exists y \forall z (z \in y \leftrightarrow (z \in x \wedge Xz))$. In less technical language: If x appears at stage α, then every member of x appears prior to α. So, given any members of x, there is a stage prior to which they all appear. And thus, given any members of x, there is a set consisting of exactly those members of x. (Remember that to establish the existence of a set, we need only show that there is a stage prior to which all the members of that set appear.)

We now know that, given any sets x and y, their intersection exists. For, by Theorem 9, we may suppose that $\forall w(w \in z \leftrightarrow (z \in x \land z \in y))$ — that is, we can let z be the set of all sets which are members of both x and y. Then $z = (x \cap y)$. More importantly, given any set x, we now know that the intersection of all the members of x (that is, $\cap x$) exists. For, by Theorems 6 and 9, we may suppose that $\forall z(z \in y \leftrightarrow (z \in \cup x \land \forall w(w \in x \rightarrow z \in w)))$ — that is, we can let y be the set of all members of members of x which are members of every member of x. Then $y = \cap x$.

We shall now establish that, given any set x, there is a set consisting of all the subsets of x. We let '$z \subset x$' mean that z is a subset of x — that is, every member of z is also a member of x (that is, $\forall w(w \in z \rightarrow w \in x)$).

Theorem 10 (Power Set): $\forall x \exists y \forall z(z \in y \leftrightarrow z \subset x)$.

Proof: Suppose that x appears at stage α and that z is a subset of x. Then every member of z appears prior to α and, thus, z cannot appear later than α (since every set appears at the first stage prior to which all its members appear). So z appears prior to the successor of α. It follows that there is a stage (namely, the successor of α) prior to which every member of the set of all subsets of x appears. So the set of all subsets of x (known as the "power set" of x) exists.

We can now show that the set of all finite Mirimanoff ordinals (that is, Mirimanoff's ω) exists. When we say that all the finite (Mirimanoff) ordinals are members of some set x, we mean that $\varnothing$ is a member of x and x is closed under $\mathfrak{F}$:

$$\varnothing \in x \land \forall y(y \in x \rightarrow \mathfrak{F}(y) \in x)$$

or, in abbreviated notation, $\Omega(x)$. When we say that all *and only* the finite ordinals are members of ω, we mean that ω is the "smallest" set which includes all the finite ordinals — that is, all the finite ordinals belong to ω and, further, ω is a subset of every set to which all the finite ordinals belong:

$$\Omega(\omega) \land \forall x(\Omega(x) \rightarrow \omega \subset x).$$

We shall now show that such an ω exists. By Theorem 8, suppose that $\Omega(x)$. By Theorem 10, let $\mathcal{P}(x)$ be the power set of x. And, by

128 Iterative Hierarchies

Theorem 9, let v be the set of all members z of $\mathcal{P}(x)$ such that $\Omega(z)$. I claim that $\cap v = \omega$ – that is, $\Omega(\cap v)$ and $\forall z(\Omega(z) \to \cap v \subset z)$.

Lemma: $\Omega(\cap v)$.

Proof: If $z \in v$, then $\Omega(z)$. And if $\Omega(z)$, then $\varnothing \in z$. So $\varnothing$ is a member of every member of v and, thus, is a member of $\cap v$. Now suppose that $w \in \cap v$. Then w is a member of every member of v. But then, since every member of v is closed under $\mathcal{F}$, $\mathcal{F}(w)$ is a member of every member of v and, thus, is a member of $\cap v$.

Lemma: $\forall z(\Omega(z) \to \cap v \subset z)$.

Proof: Suppose that $\Omega(z)$. Recall that v is the set of all subsets w of x such that $\Omega(w)$. Let v' be the set of all subsets w of $(x \cup z)$ such that $\Omega(w)$. Then $\Omega(\cap v')$ (for the reasons given in the first lemma) and $\cap v' \subset x$ (since there is a subset w of x such that $\Omega(w)$ and every member of $\cap v'$ is a member of every such subset w). So $\cap v \subset \cap v'$ (since every member of $\cap v$ is a member of every subset w of x such that $\Omega(w)$). But $\cap v' \subset z$ (for much the same reason that $\cap v' \subset x$) and, hence, $\cap v \subset z$.

These two lemmas establish that $\cap v = \omega$. We can now continue the sequence of Mirimanoff ordinals by starting with ω and repeatedly applying the operation $\mathcal{F}$. Then, starting from the beginning, we get the sequence:

$$\varnothing, \mathcal{F}(\varnothing), \mathcal{F}(\mathcal{F}(\varnothing)), \ldots ; \omega, \mathcal{F}(\omega), \mathcal{F}(\mathcal{F}(\omega)), \ldots$$

It turns out that these are the *only* Mirimanoff ordinals available in our current theory T_1. For, first of all, the sequence of Mirimanoff ordinals (ordered by $\in$) is isomorphic to the sequence of stages of our iterative hierarchy (ordered by $<$). In fact, the relation $Fx\alpha$ ("x appears at stage α") forms just such an isomorphism. This is because:

1. $\varnothing$ appears at the very first stage;
2. if a Mirimanoff ordinal x appears at stage α, then $\mathcal{F}(x)$ appears at the stage immediately succeeding α;
3. if α is a limit ordinal, then the set of all the Mirimanoff ordinals which appear at stages prior to α appears at α – and this set is itself a Mirimanoff ordinal;
4. at most one Mirimanoff ordinal can appear at each stage;
5. it follows from 1, 2, and 3 that at least one Mirimanoff ordinal appears at each stage;
6. every Mirimanoff ordinal appears at exactly one stage;
7. it follows from 4, 5, and 6 that F forms a pairing between the stages of the hierarchy and the Mirimanoff ordinals;

8. if x and y are Mirimanoff ordinals, then $x \in y$ if and only if x appears earlier than y.

7 and 8 say that F forms an order preserving pairing (that is, an isomorphism) between the stages of the iterative hierarchy and the Mirimanoff ordinals. Now recall what we know about the stages: they are well-ordered by $<$; every stage has a successor; and there is a limit ordinal. This information alone does not allow us to establish that our hierarchy of stages has a structure any more complicated than $\omega+\omega$ — that is, a structure of the form:

$$0,1,2,\ldots;\omega,\omega+1,\omega+2,\ldots$$

So, since the total sequence of our Mirimanoff ordinals is isomorphic to the overall structure of our hierarchy of stages, the Mirimanoff ordinals cited above are the only ones currently available to us. That is, the ordinals whose existence is demonstrable are all and only those which are less complex than $\omega+\omega$. (Only the totality of our ordinals is as complex as $\omega+\omega$. No individual ordinal is.)

This is very unfortunate. A theory of ordinals is meant to provide abstract representations of all possible well-orderings. And either the stages of our hierarchy themselves or the Mirimanoff ordinals which appear within those stages might seem to provide us with a marvelously elegant theory of this sort. Unfortunately, in T_1 it is easy to define well-orderings which are as complex as or even more complex than $\omega+\omega$. For example, we could order the finite Mirimanoff ordinals in a manner corresponding to the sequence

$$0,2,4,6,\ldots;1,3,5,7,\ldots$$

(which has order type $\omega+\omega$) or to the sequence

$$0,3,6,9,\ldots;1,4,7,10,\ldots;2,5,8,11,\ldots$$

(which has order type $\omega+\omega+\omega$). So some well-orderings definable in T_1 are not isomorphic either to an initial segment of the iterative hierarchy or to a Mirimanoff ordinal — indeed, some such well-orderings are more complex than the entire sequence of stages and the totality of Mirimanoff ordinals. So some available well-orderings fail to be represented within either of our theories of ordinals. And this means that these theories are of only limited value.

5. Theory T_2

We have discovered that a hierarchy of height $\omega+\omega$ will contain well-ordered structures more complex than $\omega+\omega$. We shall try to set matters aright by iterating the process of set formation through larger ordinals. It turns out to be convenient to allow our ordinals or stages to be members of sets which appear within the hierarchy. Probably the simplest way to accomplish this is to suppose that each stage α appears at the αth stage of the hierarchy:

Axiom 3.1: $\forall \alpha \, F\alpha\alpha$.

Now that our stages appear within the hierarchy, they too will be "gathered together" into sets. That is, in addition to sets of sets, there will be sets of stages and sets of both stages and sets. So we need to modify our view about where exactly in the hierarchy particular sets appear. In T_1, a set appears at the first stage prior to which all the sets belonging to it appear. We shall now say that a set appears at the first stage prior to which all the sets *and stages* belonging to it appear:

Axiom 3.2: $\forall x \forall \alpha (Fx\alpha \leftrightarrow (\forall y \in x \exists \beta < \alpha \, Fy\beta \wedge \forall \gamma \in x \exists \beta < \alpha \, F\gamma\beta \wedge \forall \beta < \alpha (\exists y \in x \forall \gamma < \beta \, \neg Fy\gamma \vee \exists \gamma \in x \forall \gamma' < \beta \, \neg F\gamma\gamma')))$.

We also adopt two axioms which make essentially the same claims about stages that Axioms 2.2 and 2.3 made about sets:

Axiom 3.3: $\forall x \exists X \exists \alpha \forall \beta (\beta \in x \leftrightarrow (X\beta \wedge \beta < \alpha))$.

Axiom 3.4: $\forall X \forall \alpha \exists x \forall \beta (\beta \in x \leftrightarrow (X\beta \wedge \beta < \alpha))$.

Axiom 3.3 says that, given any set x, either no stages at all belong to x or there are stages prior to some stage α such that all and only those stages are members of x. This corresponds to the picture that the stages belonging to a set are selected from some initial segment of the iterative hierarchy. Axiom 3.4 says that, given any stages prior to some stage α, there is a set x such that those and only those stages are members of x. This corresponds to the picture that each collection of stages from an initial segment of the hierarchy is gathered into a set.

Now that stages as well as sets are allowed to be members of sets, we need to modify our extensionality axiom—lest sets which are extensionally equivalent with respect to sets but not with respect to stages are identified. Axiom 2.4 said that sets which have exactly the same *sets* as members are identical. We now say that sets which have exactly the same sets *and stages* as members are identical:

Axiom 3.5: $\forall x \forall y ((\forall z (z \in x \leftrightarrow z \in y) \wedge \forall \alpha (\alpha \in x \leftrightarrow \alpha \in y)) \rightarrow x = y)$.

T_2 will also contain Axioms 1.1, 1.2, 1.3, 2.2, and 2.3. This means that: our stages are well-ordered by < (Axiom 1.1); some stage is a limit ordinal (Axiom 1.2); every stage has a successor (Axiom 1.3); given any set x, either x is empty or there are sets appearing prior to some stage α such that they are the only sets belonging to x (Axiom 2.2); and given any sets appearing prior to some stage α, there is a set x such that those "pre-α" sets are the only sets belonging to x (Axiom 2.3). This allows us to prove Theorem 5 (the pairing theorem). And this, in turn, guarantees that, given any sets x and y, the set $\{\{x\},\{x,y\}\}$ (designated, more briefly, by '$<x,y>$') exists. The significance of these sets and of the following theorem about them will be discussed in a moment.

Theorem 11: $\forall x \forall y \forall z \forall w(\{\{x\},\{x,y\}\}=\{\{z\},\{z,w\}\} \rightarrow (x=z \wedge y=w))$.

Proof: Suppose $\{\{x\},\{x,y\}\}=\{\{z\},\{z,w\}\}$. Then, since $\{x\}\in\{\{x\},\{x,y\}\}$ and, hence, $\{x\}\in\{\{z\},\{z,w\}\}$, either $\{x\}=\{z\}$ or $\{x\}=\{z,w\}$. That is, either $x=z$ or $x=z=w$. So $x=z$. But then $\{\{x\},\{x,y\}\}=\{\{x\},\{x,w\}\}$. It follows from the lemma given just below that $\{x,y\}=\{x,w\}$. So, again by the lemma, $y=w$.

Lemma: $\forall x \forall y \forall z(\{x,y\}=\{x,z\} \rightarrow y=z)$.

Proof: Suppose $\{x,y\}=\{x,z\}$. Then either $y=x$ or $y=z$, since $y\in\{x,y\}$. And either $z=x$ or $z=y$, since $z\in\{x,z\}$. So if $y\neq z$, then $y=x$ and $z=x$ and, hence, $y=z$. So $y=z$.

By '$x\in<x,y>$' we shall mean that $\{x\}\in\{\{x\},\{x,y\}\}$. And by '$y\in<x,y>$' we shall mean that $\{x,y\}\in\{\{x\},\{x,y\}\}$. We say that x is the *first* element of $<x,y>$ and y is the *second* element. So Theorem 11 says that sets of the form $<x,y>$ are identical only if they agree in both their first and second elements. In other words, the identity of a set of the form $<x,y>$ is determined not only by the fact that x and y are its elements, but also by the *ordering* of these elements – that is, by the fact that x is the first element and y the second. This means that, as Kazimierz Kuratowski first pointed out, the *unordered* pair $\{\{x\},\{x,y\}\}$ has all the essential properties of the *ordered* pair whose first element is x and whose second is y.[92] So, just as we can use, say, the Zermelo numbers in place of the "real" natural numbers (that is, just as we can use the Zermelo numbers to provide an interpretation of ordinary arithmetic), so we can use Kuratowski's ordered pairs in place of "real" ordered pairs (that is, we can use Kuratowski's ordered pairs to provide an interpretation of theories in which ordered pairs are treated as primitive, *sui generis* entities). Further, Kuratowski's technique provides us with representatives not only of ordered *pairs*, but also of ordered *n*-tuples for any finite *n*. For example, we can use the Kuratowski pair $<x,<y,z>>$ to represent the ordered triple

$<x,y,z>$. And we can then use $<x, <y,z,w>\!\!>$ to represent the ordered quadruple $<x,y,z,w>$. And so on.

So far, we have discussed only ordered n-tuples of sets. (For it is sets that the variables 'x', 'y', 'z', 'w' are meant to range over.) But it is easy to see that, in T_2 (which allows stages to be members of sets), we can also establish the existence both of ordered n-tuples of stages and of "mixed" ordered n-tuples involving both sets and stages. The availability of all these ordered n-tuples is of considerable importance to us because we have restricted ourselves to that branch of higher order logic which can be interpreted in the way suggested by George Boolos—namely, to *monadic* second order logic. But now, given ordered n-tuples, we can make our monadic second order variables behave like n-ary ones for any finite n. In particular, we can now both state the following axiom and put it to good use. (We could have *stated* the unabbreviated version of this axiom even without having established that $<x,\alpha>$ exists for every x and α. But in that case the axiom would not have been much use to us.)

Axiom 3.6: $\forall X(\forall y \exists!\gamma\, X<y,\gamma> \rightarrow \forall x \exists \beta \forall \gamma (\exists y \in x\, X<y,\gamma> \rightarrow \gamma < \beta))$.

'$\exists!\gamma$' means "there is exactly one stage γ such that ..." So Axiom 3.6 says that if X is a function which assigns stages to sets, then, given any set x, there is a stage β such that only stages prior to β are assigned by X to the members of x. Thus, if the members of any set are projected onto the sequence of stages, the image which is thus formed will be bounded from above by some stage—that is, there are enough stages that this image cannot be scattered throughout the whole sequence. This axiom is meant to guarantee, among other things, that no well-ordered set theoretic structure definable in T_2 can exceed in complexity every initial segment of the hierarchy of stages. So each well-ordering which can be represented by a set of T_2 is intended to be isomorphic to some initial segment of the hierarchy. Does Axiom 3.6 actually allow us to prove this?

Theorems 5, 6, 7, 8, 9, and 10 (Pairing, Union, Empty Set, Infinity, Separation, and Power Set) are provable in T_2. And so is the following useful theorem:

Theorem 12 (Replacement): $\forall X(\forall y \exists!y'\, X<y,y'> \rightarrow \forall x \exists z\, \forall y'\, (y' \in z \leftrightarrow \exists y \in x\, X<y,y'>))$.

Proof: Suppose that $\forall y \exists!y'\, X<y,y'>$. Then $\forall y \exists!\gamma \exists y'(X<y,y'> \wedge Fy'\gamma)$. So, by Axiom 3.6, pick a β such that $\forall \gamma (\exists y \in x \exists y'(X<y,y'> \wedge Fy'\gamma) \rightarrow \gamma < \beta)$. By Axiom 2.3, pick a z such that $\forall y'(y' \in z \leftrightarrow (\exists y \in x\, X<y,y'> \wedge \exists \gamma < \beta\, Fy'\gamma))$. Then $\forall y'(y' \in z \leftrightarrow \exists y \in x\, X<y,y'>)$. In less technical language: If f is a function which assigns sets to sets, then we can define a function f' which assigns to each set y the unique stage at which $f(y)$ appears. By Axiom 3.6, the image of a set x which f' projects onto the sequence of stages must be bounded by some stage β. So there is a stage prior to which all the members of the set $\{f(y):y\in x\}$ appear. Hence this set exists.

Theorem 12 says that every function from sets to sets projects the interior of a set of sets into the interior of yet another set – that is, each image created in this way will itself be a set. With the help of this theorem, we can prove that each well-ordering which can be represented by a set of T_2 is isomorphic to a Mirimanoff ordinal. So these ordinals adequately represent all available well-ordered structures. Furthermore, the relation F forms an isomorphism between the Mirimanoff ordinals and the stages of the iterative hierarchy. So these stages too could serve as the basis for an adequate theory of well-orderings.

6. A Modification of T_2

Now that we allow stages to appear within the iterative hierarchy and to be members of sets, there seems to be little reason to treat stages as non-sets. This suggests the following modification of T_2. We shall drop the special variables we have been using to denote our stages and shall add the unary relation symbol 'ON' which is to stand for the property of being an ordinal. It will generally be convenient to write '$x \in$ ON' ("x is a member of the class of ordinals") in place of 'ON(x)' ("x is an ordinal"). This notational convenience should not be taken to imply that ON is a set. I shall abbreviate strings of quantifiers by, for example, writing '$\forall x,y,z \in$ ON' in place of '$\forall x \in$ ON $\forall y \in$ ON $\forall z \in$ ON'.

Axiom 1.1':
1.11' $\forall x \in$ ON $x \notin x$
1.12' $\forall x,y,z \in$ ON$((x \in y \wedge y \in z) \rightarrow x \in z)$
1.13' $\forall x,y \in$ ON$(x \in y \vee y \in x \vee x=y)$
1.14' $\forall X(\exists x \in$ ON $Xx \rightarrow \exists x \in$ ON$(Xx \wedge \neg \exists y \in$ ON$(Xy \wedge y \in x)))$.

Axiom 1.2': $\exists x \in$ ON$(\exists y\ y \in x \wedge \forall y \in x \exists y' \in x\ y \in y')$.

Axiom 1.3': $\forall x \in$ ON$\exists y \in$ ON $x \in y$.

Axiom 1.4': $\forall x \in$ ON$\forall y \in x\ y \in$ ON.

134 ITERATIVE HIERARCHIES

These axioms say that: the ordinals are well-ordered $\in$; there is a limit ordinal; every ordinal has a successor; and every member of an ordinal is itself an ordinal.

Axiom 2.1': $\forall x \forall y (Fxy \leftrightarrow (y \in \text{ON} \land \forall z \in x \exists w \in y \ Fzw \land \forall w \in y \exists z \in x \forall v \in w \ \neg Fzv))$.

Axiom 2.2': $\forall x \exists X \exists y \forall z (z \in x \leftrightarrow (Xz \land \exists w \in y \ Fzw))$.

Axiom 2.3': $\forall X \forall y \exists x \forall z (z \in x \leftrightarrow (Xz \land \exists w \in y \ Fzw))$.

Axiom 2.4: $\forall x \forall y (\forall z (z \in x \leftrightarrow z \in y) \rightarrow x = y)$.

Axiom 3.6': $\forall X (\forall y \exists ! y' \ X <y, y'> \rightarrow \forall x \exists z \in \text{ON} \forall w \in \text{ON} (\exists y \in x \ X <y, w> \leftrightarrow w \in z))$.

The basic meaning of these axioms should be familiar since (with the exception of Axiom 2.4, which we adopt unchanged) they are merely the result of modifying some of our old axioms in order to bring them into line with our new conception of the stages of our hierarchy. Let's call this new system "T_2'."

We can now show that each ordinal x appears at stage x:

Theorem 13: $\forall x \in \text{ON} \ Fxx$.

Proof: By Axiom 1.14', let x be the first ordinal such that $\neg Fxx$. Then $\forall y \in x (y \in \text{ON} \rightarrow Fyy)$. So, by Axiom 1.4', $\forall y \in x \ Fyy$. We see immediately, then, that every member of x appears prior to x – that is, $\forall y \in x \exists z \in x \ Fyz$. Furthermore, there is no stage earlier than x prior to which all the members of x appear – that is, $\forall y \in x \exists z \in x \forall w \in z \ \neg Fyw$. For if $y \in x$, then Fyy. And, by Axioms 1.11' and 1.4', $y \notin y$. So, by Theorem 2 (which is provable in T_2'), $\forall w \in y \ \neg Fyw$. It follows, by Axiom 2.1', that Fxx – contrary to our original assumption about x. So $\forall x \in \text{ON} \ Fxx$.

Theorems 1 through 12 (*mutatis mutandis*) are all provable in T_2'. So we can once again prove that every well-ordering which can be represented by a set is isomorphic to a Mirimanoff ordinal and that F forms an isomorphism between the Mirimanoff ordinals and the ordinals of ON. It would tidy up our universe considerably if we were able to prove that the class of Mirimanoff ordinals is actually *identical* to the class ON. The following theorems will help us do so.

Theorem 14: $\forall x \in \text{ON} \forall y \in x \ y \subset x$.

Proof: Suppose $y \in x \in \text{ON}$. Then, by Axiom 1.4', $y \in \text{ON}$. And so, by Axiom 1.4' again, if $z \in y$, then $z \in \text{ON}$. Hence, by Axiom 1.12', $\forall z \in y \ z \in x$ – that is, $y \subset x$.

Theorem 15: $\forall x (\forall y \in x (y \subset x \land y \in \mathrm{ON}) \to x \in \mathrm{ON})$.

Proof: Suppose that $\forall y \in x (y \subset x \land y \in \mathrm{ON})$. And, by Theorem 3, suppose that Fxx'. Then, by Axiom 2.1', $x' \in \mathrm{ON}$ and $\forall z \in x \exists w \in x' Fzw$. But, by Theorem 13, $\forall z \in x\ Fzz$ (since $\forall z \in x\ z \in \mathrm{ON}$). So, by Theorem 2, $\forall z \in x\ \exists w \in x'\ z = w$ – that is, $\forall z \in x\ z \in x'$. Further, by Axiom 2.1', $\forall w \in x' \exists z \in x \forall v \in w\ \neg Fzv$. So, by Theorems 2 and 13, $\forall w \in x' \exists z \in x \forall v \in w\ z \neq v$. Hence, by Axioms 1.13' and 1.4', $\forall w \in x'\ \exists z \in x (w \in z \lor w = z)$. But if $w \in z \in x$, then $w \in x$. So $\forall w \in x'\ w \in x$. And hence, by Axiom 2.4, $x = x'$. So $x \in \mathrm{ON}$.

Let us now introduce a new unary relation symbol 'ORD' such that $\mathrm{ORD}(x)$ if and only if the members of x are well-ordered by $\in$ and, further, every member of x is also a subset of x. As in the case of 'ON', it will be convenient to write '$x \in \mathrm{ORD}$' in place of '$\mathrm{ORD}(x)$'. But, again, this should not be taken to mean that ORD is a set. (Otherwise our system would be open to the Burali-Forti paradox.) The sets in ORD are none other than the Mirimanoff ordinals. Indeed, the Mirimanoff ordinals are usually *defined* to be the sets in ORD. For it turns out that these sets correspond to the intuitive picture of the Mirimanoff ordinals given above (in which, starting from $\varnothing$, we form successor ordinals by applying the operation $\mathfrak{F}$ and form limit ordinals by taking the union of infinite sequences of ordinals). We shall now prove that ORD and ON are coextensive.

Theorem 16: $\forall x \in \mathrm{ON}\ x \in \mathrm{ORD}$.

Proof: By Axioms 1.1' and 1.4', the members of each set in ON are well-ordered by $\in$. So, by Theorem 14, each set in ON is also in ORD.

Theorem 17: $\forall x \in \mathrm{ORD}\ x \in \mathrm{ON}$.

Proof: Though I shall not give the details here, we can prove that the sets in ORD are well-ordered by $\in$. (The definition of ORD tells us only that the *members* of each set in ORD are well-ordered by $\in$.) So let x be the $\in$-first set in ORD such that $x \notin \mathrm{ON}$. Then $\forall y \in x\ y \in \mathrm{ON}$. And, since $x \in \mathrm{ORD}, \forall y \in x\ y \subset x$. So, by Theorem 15, $x \in \mathrm{ON}$ – contrary to our assumption about x. We conclude that $x \in \mathrm{ON}$.

7. Universal Structures

A theory is *categorical* if all its models are isomorphic – if, that is, all the structures it can be taken to characterize may be mapped onto one another by pairings which preserve relative structural position. If *abstract* structures are objects which are identical whenever they are isomorphic (if abstract structures are objects whose isomorphism is a sufficient condition for their identity), then to say that a theory is categorical is to say that it characterizes exactly one abstract structure. All of the above theories are categorical "up to the height of the hierarchy": they give exhaustive structural characterizations of the contents of any given stage, but they place no upper bound on the number of stages (though they do provide a lower bound). If we wished, we could render our theories categorical by adding axioms which supply such an upper bound. For example, we could amend T_1 in such a way that the existence of exactly *one* limit ordinal is guaranteed. A proof that the resulting theory is categorical appears in Appendix I.

We know that either the Continuum Hypothesis or its negation is a theorem of our theories because the truth or falsity of CH depends on the "breadth" rather than the height of the hierarchy. The relation between $\aleph_1$ and $2^{\aleph_0}$ is determined by the internal structure of our stages: by the strength of the "set formation" operation which carries us from stage to stage. Given a suitable formulation of CH, the number of times the "set formation" operation is iterated is completely irrelevant.[93] And, given any reasonable formulation, the truth value of CH is fixed by the contents of an initial segment of our hierarchies. (By stage ω, sets of cardinality $\aleph_0$ appear. By stage $\omega+1$, sets of cardinality $2^{\aleph_0}$ appear. By stage $\omega+3$, the pairing functions necessary for the truth of CH will have appeared – if, that is, they are going to appear at all.)

In chapter 5, we discussed "universal theories": theories within which all mainstream, formal mathematical theories are interpretable. We have now been examining "universal structures": structures in terms of which all currently orthodox mathematical structures are definable. A theory Σ can be said to capture adequately the universality of a structure only if all faithful formal characterizations of orthodox mathematical structures are interpretable within Σ. (By a "faithful formal characterization of a structure" I mean a formal theory one of whose models is that structure.) So, if all mainstream, formal mathematical theories are faithful formal characterizations of orthodox mathematical structures, it follows that all adequate theories of universal structures are universal theories. Conversely, universal theories can be construed as theories of universal structures. So, if we were to take the faithfulness of mainstream axiomatizations for granted, we would conclude that our last few chapters have dealt with two different *philosophical accounts* of a single class of theories, rather than with two different classes of theories (although one's choice of a particular sort of axiomatization may be guided by one's choice of philosophical account).

If our description of an iterative hierarchy is to be a comprehensive theory of mathematical structures, then we should demand that it provide us with a general theory of well-orderings (since these are structures of considerable mathematical interest). So the availability of elegant and comprehensive accounts of ordinals within T_2 and T_2' is a good ground for preferring these theories to T_1. Furthermore, T_2' seems preferable to T_2 because the former dispenses with T_2's domain of apparently *sui generis* ordinals whose work can be performed by the Mirimanoff ordinals.

Axiom 2.4 (Extensionality) and Theorems 5, 6, 8, 9, and 10 (Pairing, Union, Infinity, Separation, and Power Set) are the axioms of second order Zermelo Set Theory (Z^2). If we replace Separation by Replacement (Theorem 14) and add the Regularity theorem

$$\forall x(\exists y \ y \in x \rightarrow \exists y \in x \forall z \in x \ z \notin y)$$

(which is provable in T_1, T_2, and T_2' and which says that every nonempty set has an $\in$-least member) we get (second order) Zermelo-Fraenkel Set Theory (ZF^2). We have seen, then, that Z^2, but not ZF^2, is derivable in T_1, whereas both Z^2 and ZF^2 are derivable in T_2 and T_2'. T_1 is essentially a second order version of a first order theory proposed by George Boolos (although T_1 does differ from Boolos' theory in ways which are independent of the strengthening of the underlying logic).[94] Further, Boolos pointed out that axioms like our 3.6 and 3.6' allow the derivation of Replacement. Dana Scott has proposed a theory in which the stages of an iterative hierarchy are themselves sets.[95] So, in this important respect, our T_2' resembles Scott's theory. James Van Aken has recently unveiled an extremely elegant axiomatization of this type.[96] Our T_2' can be simplified using Van Aken's techniques – as I show in Appendix II.

Our treatment of iterative hierarchies in this chapter has been largely technical. In our next chapter, we shall focus on philosophical issues. The following questions will be of particular interest: Are iterative hierarchies *abstract* structures? Is it possible for isomorphic hierarchies to be distinct? When we have identified the location of a set within an iterative hierarchy, do we need to say anything more to specify *which* set occupies that location? Is there any more which *can* be said?

IX
STRUCTURALISM

> Mathematicians do not study objects, but the relations between objects; to them it is a matter of indifference if these objects are replaced by others, provided that the relations do not change. Matter does not engage their attention; they are interested by form alone. — Henri Poincaré[97]

> ... what matters in mathematics ... is not the intrinsic nature of our terms, but the logical nature of their interrelations. — Bertrand Russell[98]

1. Stenius on Sets and Structures

We shall begin our discussion of structuralism by picking up a thread from chapter 3. Max Black's essay "The Elusiveness of Sets" elicited a rich and thoughtful response from Erik Stenius. Recall Black's contention that one's mastery of everyday set or plurality talk is sufficient to provide one with an essentially sound notion of mathematical set. In his essay "Sets," Stenius disputes this claim and proposes a very different strategy for making sense of mathematical sets. He suggests that mathematical "set" theories are not in fact concerned with sets in any everyday sense, but rather are theories of abstract structures. This structuralist account of set theory is extremely compelling — even though Stenius' own arguments for it are tainted by his reliance on an overly narrow conception of inter-theoretic interpretation.

Let $\mathfrak{D}$ be some fixed universe of discourse consisting entirely of individuals (i.e., non-sets). And let L be a first order language whose pronominal expressions refer singularly to objects within $\mathfrak{D}$. Stenius notes that the addition of plurally referring expressions to L need not involve any expansion of the universe of discourse $\mathfrak{D}$. For the new plural expressions need differ from the old singular ones only in that

Stenius on Sets and Structures 139

the latter refer to individuals within 𝔇 *one at a time* while the former refer to them *many at a time*. Indeed, it is precisely the ontological inoffensiveness of plural expressions which leads Black to recruit them for his proposed account of mathematical set talk. But mightn't this very trait which makes plural expressions philosophically attractive also render them unfit to support a commonsensical explication of mathematical set theories?

Stenius asks us to consider how we might account for mathematical set talk without straying from a fixed universe of individuals 𝔇. Apparent references to sets of such individuals could be explained away as disguised plural references to those individuals themselves. But apparent references to sets of sets of individuals could not be straightforwardly treated as plural references to sets of individuals – for there are no such sets in 𝔇 to be referred to. Remember: our use of expressions which refer plurally to individuals does not require that sets of individuals be added to 𝔇 nor does it magically produce them. How then are mathematical references to sets of sets to be explained?

Stenius suggests that we must use the individuals available within 𝔇 as *representatives* of the mathematical sets we are trying to explain away – each individual representing no more than one set. If we have enough individuals to go around, this would just mean *structuring* our universe 𝔇 in a certain way. We would not have to *expand* 𝔇. Suppose, for example, that there are at least four objects in 𝔇. Then we could account for the tiny set theoretic universe

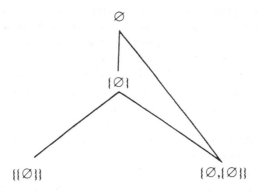

by picking four individuals a, b, c, d from 𝔇 and defining a relation $\in'$ such that $a \notin' a$, $a \in' b$, $a \notin' c$, $a \in' d$, $b \notin' a$, $b \notin' b$, $b \in' c$, $b \in' d$, $c \notin' a$, $c \notin' b$, $c \notin' c$, $c \notin' d$, $d \notin' a$, $d \notin' b$, $d \notin' c$, $d \notin' d$. We would then have defined the structure

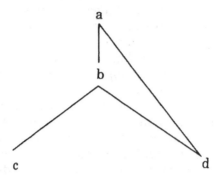

which, of course, is isomorphic to the fragment of the iterative hierarchy pictured above. We could then take, say, the assertion that the empty set is a member of its own singleton as a disguised way of asserting that the individual a stands in the relation $\in'$ to the individual b. If we were asked the number of elements in the power set of the power set of the empty set, we would count up the objects which stand in $\in'$ to d. In general, we would treat apparent references to a particular set S as references to the individual whose role in the structure determined by $\in'$ is the same as that played by S in the structure determined by $\in$.

At this point, Stenius' argument takes quite an interesting turn. To say that certain individuals of $\mathfrak{D}$ are to *represent* certain mathematical sets suggests that mathematical sets are entities in their own right which are distinguishable from the individuals standing proxy for them – or, at least, it suggests that we know what it would be like if real, honest-to-goodness sets inhabited our universe rather than mere imitations. But Max Black's chief reason for seeking a commonsensical explication of mathematical set talk is precisely our (alleged) unclarity about what mathematical sets are and whether they really deserve to be called 'sets' in some everyday sense. If Stenius is to play by Black's rules, he mustn't *presuppose* that we adequately understand the notion of mathematical set – rather he must *supply* us with such an understanding. This he attempts to do.

If we accept Stenius' explication of mathematical set talk in terms of talk about individuals, we must grant that our choice of, say, individuals a, b, c, d to play the roles of the sets $\varnothing$, $\{\varnothing\}$, $\{\{\varnothing\}\}$, and $\{\varnothing,\{\varnothing\}\}$ is to a considerable extent arbitrary. All that matters is that we be able to define the appropriate sort of relation between the chosen individuals. Any four individuals and any relation between them which form the desired sort of structure are acceptable. No model which has the structure

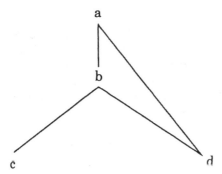

is to be regarded as inappropriate. But this willingness to accept any member from an equivalence class of isomorphic models can be regarded as the pragmatic content of the claim that the theory under consideration is a theory of an abstract structure. That is, the claim that a theory is *about a certain abstract structure* can be taken to mean that the theory is about any system of objects which *has that structure*. Any such system will then be accepted as one of the intended models of the theory. Thus if we express indifference about the nonstructural characteristics of proposed models for mathematical set theories, we seem *ipso facto* to be expressing the belief that those theories are theories of abstract structures. Stenius means to draw us toward precisely such a structuralist view.

We are in a position to rebut an obvious criticism of Stenius' explicative strategy: namely that it would allow us to make sense of very little mathematical set talk. Let us first consider how this criticism might be formulated. Stenius believes that supplying an explication of a set theory is a matter of constructing set theoretic structures within our universe of individuals. If we are to account for our apparent references to large and complex iterative hierarchies of sets in the way Stenius proposes, our universe of available individuals must be at least as large and complex as these hierarchies themselves. For remember: no more than one set is to be represented by each individual. If our universe of discourse contained only a small infinite number of individuals, we would only be able to account for rather weak and uninteresting mathematical set theories – and universal theories, in the sense of chapter 5, might be beyond our grasp.

If Stenius' primary intention were indeed to sketch out a reduction of mathematical set talk to talk about individuals, this criticism would be very much to the point. But, as we have just seen, Stenius' discussion has a very different aim. If I understand him correctly, all he really means to show is that our evaluation of proposed models for a set theory will focus on the structural properties of those models. That is, Stenius asks us to consider some interpretations of trivial set theoretic assertions not as a first step in a grand reductive program, but rather as a stage in his argument for a structuralist account of mathematical set theory. The improbability of a grand reduction does not undercut Stenius' central thesis: namely that the "nodes" in an iterative hierarchy do not come already filled up by objects which deserve to be called "sets" in any everyday sense; rather, those nodes are contentless place holders to be filled by whatever objects happen to interest us at the moment. While our commonsense talk about sets is to be explained in terms of plural reference, our mathematical set theories are to be regarded as theories of abstract structures.

I have considerable sympathy for the view that mathematics is the science of structure. However, I have some reservations about Stenius' argument for structuralism. Briefly: he has built a structuralist bias into his conception of inter-theoretic interpretation; so this conception cannot serve as a basis for a really compelling argument for a structuralist view of set theory. Stenius says in effect that an interpretation function which maps a set theory into a theory of individuals may tinker only with the symbol '$\in$' and with the range of the quantifiers, but is to leave '=' alone. That is, the allowable interpretations i are to be such that

$i(\alpha \in \beta)$	is	$\theta(\alpha,\beta)$
$i(\forall \alpha\ \varphi)$	is	$\forall \alpha(\psi \to i(\varphi))$
$i(\exists \alpha\ \varphi)$	is	$\exists \alpha(\psi \wedge i(\varphi))$
$i(\alpha=\beta)$	is	$\alpha=\beta$

where θ is some binary predicate other than '$\in$'. The reader will recall that in my discussion of interpretation and abstraction in chapter 4, I noted that interpretations which tinker with identity are likely to be resisted by mathematical structuralists. This is because such interpretations will map statements about structures of a particular sort into statements about the corresponding "pseudo-structures." For example, a statement about a dense linear ordering will be mapped into a statement about a pseudo–dense-linear-ordering – where a pseudo–dense-linear-ordering is a structure which appears to be a dense linear ordering only because some relatively weak relation is successfully impersonating identity. But this sort of translation involves a loss of precisely the content which a structuralist would regard as most important (namely – need I say it? – the structural content). Thus, by insisting that explications of set theory leave identity unmolested, Stenius restricts us to precisely those interpretation functions which are acceptable from a structuralist point of view. However, since there is no reason to think that a nonstructuralist will accept this restriction in the absence of any compelling argument for it and since no such argument is provided by Stenius, his case for structuralism can be expected to impress nonbelievers very little.

2. Reference and Reduction

A more promising argument for mathematical structuralism stems from a paper by Paul Benacerraf.[99] Since Benacerraf's own presentation is difficult to make out, we shall rely on a reformulation by Philip Kitcher.[100] Consider a theory of the natural numbers whose language contains infinitely many numerals of the usual sort: '0', '1', '2', '3', and so on. What meaning should we ascribe to these numerals? The most natural view is that each numeral is a proper name which refers to exactly one number. This certainly accords with the surface grammar of arithmetical statements: '4=5−1' seems to have the same syntactic form as 'Gottlob Frege is the author of *Begriffsschrift*', so it is tempting to suppose that '4' refers to a single specifiable object just as 'Gottlob Frege' does. Unfortunately, this jars with some equally tempting theses about the foundations of mathematics. As Kitcher remarks, the following four statements cannot all be consistently maintained:

1. Numerals refer to non-spatio-temporal objects.
2. If numerals refer to objects, then each numeral refers to exactly one object.
3. All non-spatio-temporal objects are sets.
4. Given any numeral, there is no satisfactory way to specify a unique set to which that numeral refers.

Statement 1 is a tenet of mathematical platonism: the doctrine that each meaningful mathematical sentence is either true by virtue of accurately characterizing certain non-spatio-temporal objects or false by virtue of inaccurately characterizing those objects. Mathematical platonism is firmly grounded at least to the following extent.

Once one has granted that there is a domain of objects whose properties render set theoretic theorems true, one is constrained to locate that domain outside of space-time (since, for example, an iterative hierarchy of height greater than $\omega+1$ exceeds the whole of space-time in size and complexity).

Statement 2 is a tenet of semantic unitarianism: the doctrine that syntactically similar sentences should be provided with correspondingly similar truth conditions. The basic idea is that, *ceteris paribus*, a single comprehensive semantic theory is preferable to a host of restricted ones. Since the numeral '4' and the proper name 'Gottlob Frege' occupy the same syntactic category and since 'Gottlob Frege' and other successfully referring proper names are each assigned a single object as their referent, a semantic unitarian would be strongly inclined to assign '4' a single object as its referent (once it is granted that '4' refers at all).

Statement 3 is an extreme form of set theoretic reductionism: the doctrine that theoretical unification in mathematics should be attained by construing all mathematical sentences as assertions about sets. We have seen that set theoretic reductionism figures prominently in the ideology of most practicing mathematicians. And, as we saw in chapter 1, a commitment to this doctrine can be powerfully defended. Of course, the claim that all non-spatio-temporal objects are sets is harder to swallow than the claim that all *mathematical* objects are sets. (Mightn't one suppose, for example, that Justice and Piety are nonmathematical non-sets which lie outside space-time?) Anyone who finds this sweeping reductionism unpalatable should feel free to rewrite statements 1 and 3 as follows.

1'. Numerals refer to non-spatio-temporal *mathematical* objects.
3'. All non-spatio-temporal *mathematical* objects are sets.

This will do no violence to our argument for mathematical structuralism.

Statements 1, 2, and 3 (or 1', 2, and 3') imply that each numeral refers to exactly one set. This invites the obvious question: To which sets do the numerals refer? It is important to note at this point that the existence of infinitely many sequences of order type ω can be proved in any moderately strong set theory. The numerals can be taken to refer to the elements of any such sequence without undermining the truth of any arithmetical theorem. In fact, a given numeral can safely be taken to refer to any set one wishes (since every set is the nth element of some ω-sequence for any natural number n). It is therefore occasionally suggested that no assignment of sets to numerals is any more defensible than any other. In this spirit, we might, for example, exclaim: "Let '4' denote anything you want (the empty set, the set of all denumerable ordinals, the 43rd stage of the iterative hierarchy – whatever); one assignment is exactly as good as any other!"

A practicing set theorist would regard this unbridled libertarianism as extremely naive. As far as actual mathematical practice is concerned, some assignments of sets to numerals are clearly preferable to others. Here's an example: Let Ψ be a formalization of first order logic. Let f be a function which pairs the finite Mirimanoff ordinals and the sentences of Ψ. We shall use '0*', '1*', '2*', and so on to denote the finite Mirimanoff ordinals. We now assign numerals to ordinals as follows. Let '0' denote 0* if $f(0*)$ is a theorem of Ψ and let '0' denote 1* otherwise; let '1' denote 2* if $f(1*)$ is a theorem of Ψ and let '1' denote 3* otherwise; in general, let a numeral n denote $(2n)*$ if $f(n*)$ is a theorem of Ψ and let n denote $(2n+1)*$ otherwise. Since there is no dependable mechanical procedure for determining whether an arbitrary sentence of Ψ is a theorem, there is likewise no such procedure for determining what the referent of an arbitrary numeral is under the above assignment. So if we were to adopt this assignment, we could rest assured that each numeral has a unique referent, but we would not generally be in a position to specify what that referent is. I submit that nothing short of lunacy would inspire a mathematician to adopt such a mapping of numerals into sets as a canonical representation of the natural numbers.

So it is false that "anything goes" when it comes to fixing the referents of numerals. But neither are we so tightly constrained that there is a unique "right" assignment of sets to numerals. In fact, the looseness of the theoretical constraints currently in place renders senseless any attempt to discover *the* "right" assignment. A mapping of numerals into sets cannot be "correct" or "incorrect" – it can only be more or less useful with respect to some specified mathematical project. For example, within some theories, arithmetic is most conveniently developed using the Zermelo numbers $(\varnothing, \{\varnothing\}, \{\{\varnothing\}\}, \ldots)$. On the other hand, if one's primary concern is the general theory of ordinals (both finite and infinite), then it is often best to treat the finite Mirimanoff ordinals as one's natural numbers. The question of whether the Zermelo or Mirimanoff numbers are the "real" natural numbers would be regarded as bizarre and pointless by a mathematician.

According to Statement 4, there is no satisfactory way to specify a unique set to which a given numeral refers. Of course, there are any number of ways of doing such specifying which are satisfactory *with respect to a particular mathematical project.* But when mathematicians fix the referent of a numeral they reserve for themselves the right to switch things around when their interests change. So, from the point of view of a practicing mathematician, there is likely to be no satisfactory way of assigning a set to a numeral *once and for all.* And a semantic account of arithmetic which assigns each numeral a fixed set would seriously misrepresent the actual way in which mathematicians treat arithmetical languages. Having once granted that philosophers of mathematics should be guided by the actual practice of mathematicians, such a misrepresentation becomes intolerable.

We are thus faced with a dilemma. Statements 1, 2, 3, and 4 (or 1', 2, 3', and 4) are all plausible theses about the foundations of arithmetic. But 1, 2, and 3 (or 1', 2, and 3') imply the falsity of 4. So we cannot consistently assert *all* of these statements. Which, then, should we abandon? Paul Benacerraf offers arguments which undermine Statement 2: he supplies us with reasons for thinking that numerals, though they might perhaps be taken to refer to objects, do not each refer to a *single* object.

> ... there is no more reason to identify any individual number with any one particular object than with any other ... For arithmetical purposes the properties of numbers which do not stem from the relations they bear to one another in virtue of being arranged in a progression are of no consequence whatsoever. But it would be only these properties that would single out a number as this object or that ... [I]n giving the properties ... of numbers you merely characterize an *abstract structure* ... [T]he "elements" of the structure have no properties other than those relating them to other "elements" of the same structure. If we identify an abstract structure with a system of relations [we find that arithmetic elaborates] the properties of the "less-than" relation, or of all systems of objects ... exhibiting that abstract structure.[101]

Referentially successful proper names contribute to the meaning of the sentences in which they occur by allowing those sentences to characterize particular specifiable objects (the referents of the proper names). Numerals, according to Benacerraf, contribute to the meaning of the sentences in which they occur by allowing those sentences to characterize the structural features shared by every ω-sequence. A numeral is able to play this role precisely because it does not refer to any one object. So Benacerraf leads us to conclude that Statement 2 is either false or (if one wishes to say that numerals do not refer at all) only vacuously true. Benacerraf himself concludes, rather mysteriously, that "numbers could not be objects." To this I can only respond that Benacerraf's argument is considerably more lucid than the conclusion it allegedly supports.

Richard Dedekind proposed a structuralist account of arithmetic nearly eighty years prior to the appearance of Benacerraf's paper. Dedekind agrees with Benacerraf that numerals contribute to the meaning of arithmetical sentences by allowing them to characterize the invariant structural features of ω-sequences. But Dedekind nonetheless seems to believe that each numeral refers to exactly one object (albeit a most unusual object). He writes: "If in the consideration of a simply infinite system N [whose members are arranged in an ω-sequence by a relation <] we entirely neglect the special character of the elements; simply retaining their distinguishability and taking into account only the relations to one another in which they are placed by [<], then are these elements called *natural numbers*.... With reference to this freeing the elements from every other content (abstraction) we are justified in calling numbers a *free creation of the human mind*."[102] On this view, numerals refer to positions in the unique abstract structure of order type ω. Let's see what this is supposed to mean.

Dedekind asks us, first, to imagine an ω-sequence δ whose elements can be identified independently of their placement in δ. δ is a *concrete* structure of order type ω. We are now to perform a mental act of abstraction in which we "free" the elements of δ from every feature which serves to distinguish δ from any other ω-sequence. This process generates an *abstract* ω-sequence δ'. In fact, since we have rendered δ' indiscernible from any "other" abstract ω-sequence (having abstracted from those features which make discrimination possible), we may speak of *the* (one and only) ω-sequence δ'. Each numeral, then, refers to a particular element of δ'. Elements of δ' are radically denatured counterparts of the elements of δ. Indeed, they are nothing more than ghostly place-holders or contentless positions. If a contentless position fails to live up to your notion of an *object*, then feel free to deny that numerals refer to objects. The point still stands, however, that each numeral refers to exactly one call-it-what-you-will.

Perhaps I should emphasize that this is Dedekind's view – not mine. I regard such talk of "creating objects via mental acts of abstraction" as mere metaphor. But (and *this* I should certainly emphasize) it is a highly suggestive form of metaphor. We shall now consider a promising attempt at turning Dedekind's philosophically rich poetry into philosophically defensible prose. We shall also see how arithmetical structuralism can be extended to *set theoretic* structuralism:

i.e., the view that mathematical sets are mere positions in abstract structures.

3. Resnik's Mathematical Structuralism

So far, we have seen why *arithmetic* ought to be regarded as a characterization of an abstract structure. In a path-breaking series of essays, Michael Resnik has ventured a Dedekind-like account of mathematics as a whole — set theory included. Resnik proclaims: "The objects of mathematics ... are structureless points or positions in structures. As positions in structures, they have no identity or features outside of a structure."[103] In particular: "... the set theoretic hierarchy ... is a pattern [i.e., an abstract structure]. Pure sets are positions in it whose apparent internal structures are a fabrication of their relationships to other positions in the hierarchy."[104] Faced with the question of what abstract structures are, he writes that he is "... inclined to reject any solution that defines [abstract structure] in set theoretic terms because ... we may achieve a clearer understanding of sets by thinking of them as nodes in a certain kind of [abstract structure]."[105]

Resnik develops two main arguments for his structuralism. First, he maintains that his position allows us to resolve the Benacerraf-Kitcher dilemma. Second, he maintains that mathematical structuralism promises to allay epistemological doubts about the very possibility of mathematical knowledge. As to this latter claim, I refer the reader to Resnik's own exposition.[106] We shall concentrate entirely on the first argument.

Resnik's version of the Benacerraf-Kitcher dilemma is refreshingly brief: "... mathematical theories are incapable of distinguishing their isomorphic models.... According to the Platonist numerals, functional constants, set abstracts, etc. refer to particular mathematical objects, but from what I have said about isomorphic models, it appears that not a single one of these is identifiable absolutely."[107] Again: "... no mathematical theory can do more than determine its objects up to isomorphism. Thus the platonist seems to be in the paradoxical position of claiming that a given mathematical theory is about certain things and yet be unable to make any definitive statement of what these things are."[108] What we have here is the scheme of a cogent argument. But Resnik has left some very important gap-filling up to us.

Let T be a categorical set theory whose only non-logical symbol is '∈'. Resnik takes it for granted that T's truth conditions are satisfied by any model having the right structure (the particular contents of the model being of no significance). But this will be so only if T's truth conditions are fixed entirely by T's logical vocabulary. If we are able to attach some determinate extralogical significance to '∈', this will restrict the contents of possible models of T. (There is nothing else in which the determinacy of an interpretation of '∈' could consist.) So, in order to save Resnik's argument, we must show that no viable interpretation of this sort is available. If we ignore this task we cannot claim to have argued effectively for set theoretic structuralism. We shall, in fact, have done little more than *presuppose* it.

We now have a better feeling for what was at stake in chapters 2 and 3. We there made significant progress toward establishing that an appropriate extralogical content for '∈' is to be found neither in the history of set theory nor in ordinary language set talk. If these twin projects could be completed in full detail, it would render set theoretic structuralism unavoidable – at least among philosophers who are both mathematical platonists and mainstream semantic unitarians (a *mainstream* semantic unitarian being someone who embraces a standard semantics for first order logic).

Suppose it is a theorem of T that $\exists!x \neg \exists y\ y \in x$ ("there is exactly one empty set"). A mathematical platonist would say that this theorem is true by virtue of accurately characterizing some domain of non-spatio-temporal objects. A mainstream semantic unitarian would say that the truth of this theorem implies that the intended quantificational domain features a single object $\emptyset$ which satisfies the formula '$\neg \exists y\ y \in \emptyset$' (since any other reading would saddle us with a deviant semantics). Suppose we accept each of these postulates. And suppose we admit that there is no viable interpretation of '∈' which fixes the contents of T's models. This would mean, for example, that we are able to characterize $\emptyset$ *only* by citing its position in the abstract structure determined by T. What should we conclude, then, about $\emptyset$? What sort of an object is $\emptyset$? Given an unwavering commitment to mathematical platonism and mainstream semantic unitarianism, I see no alternative to supposing that $\emptyset$'s identity is determined by and *only* by its position in the T-hierarchy – for the attribution of any other identity conditions to $\emptyset$ could not be justified by any feature of mathematical practice and, even worse, would probably create a distorted image of that practice. On this view, T

is the theory of a single abstract structure whose constituents (such as $\varnothing$) are no more than contentless nodes or positions. This is essentially Resnik's doctrine.

Perhaps I should emphasize that Resnik is indeed a mathematical platonist: "Since structures are [non-spatio-temporal] entities, structuralism is a variant of [p]latonism."[109] Furthermore, Resnik's doctrine is compatible with a form of semantic unitarianism: "Structuralism is ... at ease with a correspondence theory of truth and a referential semantics. Thus no special theory of mathematical meaning is needed to incorporate this view into a general theory of language.... A uniform theory of meaning and truth can be given for all of science including mathematics since truth in mathematics can be taken to be correspondence with mathematical reality."[110] We should now discuss whether mathematical structuralism of the Dedekind-Resnik variety is compatible with set theoretic reductionism.

First a slight digression. Let γ and γ' be the abstract structures characterized by two second order, categorical theories Γ and Γ' respectively. We say that γ *occurs in* γ' if Γ is interpretable in Γ'.[111] This means, for example, that the abstract ω-sequence (corresponding to the natural numbers ordered by "less than") occurs in the iterative hierarchy of height $\omega+\omega$ – for second order arithmetic is interpretable in the categorical extension of Z^2 which asserts the existence of exactly one limit ordinal. In fact, we might say that the abstract ω-sequence occurs in the iterative hierarchy of height $\omega+\omega$ *infinitely many times* – corresponding, for example, to the infinitely many predicates to which an interpretation function might relativize the quantifiers of Z^2. In less forbidding terminology: we can map the abstract ω-sequence onto the Zermelo numbers or onto the Zermelo numbers greater than $\varnothing$ or onto the Zermelo numbers greater than $\{\varnothing\}$ or onto the Zermelo numbers greater than $\{\{\varnothing\}\}$ or onto many other sectors of an iterative hierarchy. One thus generates infinitely many occurrences of the abstract ω-sequence.

This notion of structural occurrence is quite useful. But if it is taken in the wrong way, it can lead to needless perplexities. The notion of structural occurrence which *ought* to be at work here is analogous to the notion of the occurrence of a variable in a formula. There are, for example, three distinct occurrences of the variable 'x' in the formula '$\forall x\ x = x$'. None of these occurrences are to be identified with the variable 'x' itself. Similarly, there are infinitely many distinct occurrences of the abstract ω-sequence in the abstract iterative hierarchy of height $\omega+\omega$. None of these occurrences are to be identified with the abstract ω-sequence itself.

It is common to treat 'abstract' and 'concrete' as synonyms for 'non-spatio-temporal,' and 'spatio-temporal'.[112] But there is a perfectly respectable sense (stemming from Husserl) in which the "abstractness" of a structure consists in the contentlessness of its constituents. A structure is then to be considered "concrete" insofar as its constituents can (or, even worse, must) be characterized independently of their structural position. This is entirely consistent with Dedekind's conception of abstract structures as products of a severe abstraction process. Having made it clear that we are now speaking of Husserl-Dedekind abstractness, it's easy to see that the abstract ω-sequence cannot itself be *part* of the abstract iterative hierarchy of height $\omega+\omega$. It is essential to each element of the abstract ω-sequence that it occupy a particular position in *that* sequence; it is essential that its identity be determined by and only by its standing in certain relations to other elements of that sequence. If an ω-sequence α features an element whose identity cannot be fixed by citing its position in α or can be fixed without citing that position, then that sequence is not abstract. This is why an ω-sequence α within an abstract iterative hierarchy is invariably *concrete* – it invariably features elements whose identity cannot be fixed merely by citing locations within α but can be fixed without citing any such locations.

For example, let the ω-sequence α be a proper sub-structure of the abstract $\omega+\omega$-hierarchy (i.e., the abstract iterative hierarchy of height $\omega+\omega$). If α were itself the *abstract* ω-sequence, then we could identify the first element of α merely by noting that *it is α's first element*. As things stand, however, the first element of α can be identified only by locating it within the $\omega+\omega$-hierarchy as a whole – for example, by noting that the first element of α is that unique element of the $\omega+\omega$-hierarchy to which nothing stands in $\in$. In the absence of such a specification, the first element of α could, for all we know, be *any* element of the $\omega+\omega$-hierarchy (since any such element is fit to stand at the beginning of an ω-sequence). So, proceeding more methodically, we can argue as follows: every element of α is an element of the abstract $\omega+\omega$-hierarchy; each element of the abstract $\omega+\omega$-hierarchy can be identified *only* by citing its position in that hierarchy; we cannot adequately specify a position within the abstract $\omega+\omega$-hierarchy merely by citing a position within α; so we cannot identify any element of α merely by citing a position within α; so α is not abstract. On the other hand, we *can* identify each element of α by citing its position within the abstract $\omega+\omega$-hierarchy (regardless of whether we indicate its position within α). This is yet another reason why α is not abstract. We conclude that it is wrong to identify the abstract ω-sequence with any of its occurrences in larger abstract structures.

If one fails to keep abstract structures and their occurrences rigidly distinct, one risks being faced with a reincarnation of the Benacerraf-Kitcher dilemma. In the previous section, we raised the question of which set is the "real" referent of a numeral. We might now (perversely) raise the question of which occurrence of an abstract structure is "really" that structure. We might ask whether, say, the sequence of Zermelo numbers is "really" the abstract ω-sequence. We will then be confronted anew by the sort of problem which impelled us in the direction of structuralism in the first place. For precisely this reason, our set theoretic reductionism must not commit us to any *identification* of the abstract ω-sequence with a sub-structure of an iterative hierarchy. This just means that we must carefully distinguish between the abstract ω-sequence and its concrete occurrences.

What then becomes of set theoretic reductionism? Can we salvage any of its central features? It would be wrong (or, at least, misleading) for us to continue to insist that all meaningful mathematical sentences are assertions about sets. I would go so far as to say that a sentence is an assertion about sets when and only when it is interpreted as making some claim about an abstract iterative hierarchy whose height is some limit ordinal. (Hierarchies whose height is a successor ordinal are highly nonstandard by virtue of not being closed under the operation of forming power sets.) If a sentence is interpreted as making some claim about the abstract ω-sequence, then (when so interpreted) it is not an assertion about sets.

However, this consideration requires us to modify the reductionist program only very little. We might, for example, maintain that every meaningful mathematical sentence is an assertion about an abstract structure some of whose *occurrences* appear within iterative hierarchies. Mathematical sentences would then all get to be about sets in an only slightly extended sense. From this point of view there is nothing objectionable about using numerals to label positions in a set theoretic ω-sequence.

> Borrowing the notation and axioms of number theory to describe a portion of the [abstract] set theoretic [hierarchy] will introduce no falsehoods into set theory and it will call attention to the occurrence of the [abstract ω-]sequence in question. That there is such an occurrence is surely as interesting mathematically as it is philosophically. Bringing number theory in suggests developments in set theory itself too — transfinite generalizations of number theory. Thus the practice now

standard in set theory of using numerals to designate certain sets is justified from a pragmatic point of view.[113]

The mathematical motivation for any sort of reductionism is that it promotes theoretical unification and cross-fertilization. Mathematical structuralism can easily accommodate a version of set theoretic reductionism which contributes to the attainment of these goals. Reductionism and structuralism are not fundamentally at odds with one another.

Mathematical structuralism of the Dedekind-Resnik variety, particularly when combined with the Boolos interpretation of second order logic, promises to solve a vast range of problems in the philosophy of set theory. Let me stress, however, that this is indeed largely a matter of *promise,* not of actual achievement. For example, structuralists are often better at indicating what abstract mathematical structures are *not* than at specifying what they are. This sort of "negative theology" leads some philosophers to doubt whether the notion of structure is any clearer than the unanalyzed notion of set. Furthermore, we are still haunted by the specter of our own Dummettian argument (chap. 5, §4). Until we show that the truth conditions of set theoretic propositions can be fixed by something other than overt set theoretic proof practices, we cannot be sure that structuralism is viable (since we cannot then be sure that the standard set theories are anything other than nonsense). In this state of unresolved tension, I leave it to you, my reader, to write the next chapter.

APPENDIX I

We transform T_1 into T_{1*} by replacing Axiom 1.2 with:

Axiom 1.2*: $\exists!\alpha(\exists\beta\ \beta<\alpha \wedge \forall\beta<\alpha\exists\gamma<\alpha\ \beta<\gamma)$.

Whereas Axiom 1.2 asserts merely that there is a limit ordinal, Axiom 1.2* asserts that there is *exactly one* limit ordinal. We shall now prove that T_{1*} is categorical (i.e., that all its models are isomorphic to one another).

A model S of T_{1*} consists of a domain $\mathfrak{D}_S$ whose members serve as the sets of T_{1*}, a domain ON_S whose members serve as the ordinals of T_{1*}, a binary relation $\in_S$ whose field is $\mathfrak{D}_S$, a binary relation $<_S$ whose field is ON_S, and a binary relation F_S which maps $\mathfrak{D}_S$ into ON_S.

Let S and S' be models of T_{1*}.

Theorem 1: ON_S is well-ordered by $<_S$ and $ON_{S'}$ is well-ordered by $<_{S'}$.

Proof: Since S and S' satisfy Axioms 1.11–1.13, $<_S$ and $<_{S'}$ are irreflexive, transitive, and connected. Since S and S' satisfy Axiom 1.14, there are no members of ON_S among which there fails to occur a $<_S$-least and there are no members of $ON_{S'}$ among which there fails to occur a $<_{S'}$-least.

Theorem 2: ON_S and $ON_{S'}$ each have exactly one limit point.

Proof: Since S satisfies Axiom 1.2*, exactly one member of ON_S has a $<_S$-predecessor but no immediate $<_S$-predecessor. Since S' satisfies Axiom 1.2*, exactly one member of $ON_{S'}$ has a $<_{S'}$-predecessor but no immediate $<_{S'}$-predecessor.

Theorem 3: Each member of ON_S has a $<_S$-successor and each member $ON_{S'}$ has a $<_{S'}$-successor.

Proof: S and S' satisfy Axiom 1.3.

Now let 0_S be the $<_S$-first member of ON_S. And let $0_{S'}$ be the $<_{S'}$-

156 APPENDIX I

first member of $ON_{s'}$. Let ω_s be the limit point of ON_s. And let $\omega_{s'}$ be the limit point of $ON_{s'}$. If $\alpha \in ON_s$, let $\sigma(\alpha)$ be the $<_s$-first member of ON_s which is $<_s$-greater than α. And if $\alpha' \in ON_{s'}$, let $\sigma'(\alpha')$ be the $<_{s'}$-first member of $ON_{s'}$ which is $<_{s'}$-greater than α'.

Let the function f be defined recursively as follows:

$f(0_s) = 0_{s'}$
$f(\sigma(\alpha)) = \sigma'(f(\alpha))$
$f(\omega_s) = \omega_{s'}$.

Theorem 4: $\forall \alpha, \beta \in ON_s(f(\alpha) = f(\beta) \to \alpha = \beta)$.

Proof: Suppose $\alpha, \beta \in ON_s$ and $f(\alpha) = f(\beta)$. We must consider three cases. Case 1: $\alpha = 0_s$. Then $f(\alpha) = 0_{s'}$ and, hence, $f(\beta) = 0_{s'}$. If $\beta \neq 0_s$, then $f(\beta)$ has at least one $<_{s'}$-predecessor (since β would then be either $\omega_{s'}$ or a successor ordinal). But $0_{s'}$ has no predecessors. So $\beta = 0_s$ and, hence, $\alpha = \beta$. Case 2: $\alpha = \omega_s$. Then $f(\alpha) = \omega_{s'}$ and, hence, $f(\beta) = \omega_{s'}$. If $\beta \neq \omega_s$, then $f(\beta)$ has either an immediate $<_{s'}$-predecessor or no $<_{s'}$-predecessor at all (since β would then be either $0_{s'}$ or a successor ordinal). But $\omega_{s'}$ lacks this property. So $\beta = \omega_s$ and, hence, $\alpha = \beta$. Case 3: $\alpha = \sigma(\gamma)$. Then $f(\alpha) = \sigma'(f(\gamma))$ and, hence, $f(\beta) = \sigma'(f(\gamma))$. As an inductive hypothesis, suppose γ satisfies the theorem (i.e., given any $\delta \in ON_s$, $f(\gamma) = f(\delta)$ only if $\gamma = \delta$). If either $\beta = 0_s$ or $\beta = \omega_s$, then $f(\beta)$ has no immediate $<_{s'}$-predecessor. But $\sigma'(f(\gamma))$ does have an immediate $<_{s'}$-predecessor. So we can pick a $\delta \in ON_s$ such that $\beta = \sigma(\delta)$. Then $f(\beta) = \sigma'(f(\delta))$ and, hence, $\sigma'(f(\gamma)) = \sigma'(f(\delta))$. So $f(\gamma) = f(\delta)$ (since the function σ' is one-to-one) and, hence, $\gamma = \delta$. But then $\sigma(\gamma) = \sigma(\delta)$ and, hence, $\alpha = \beta$. We have shown that, in each of the three possible cases, $f(\alpha) = f(\beta)$ only if $\alpha = \beta$. (To say that σ' is one-to-one is to say that $\forall \alpha', \beta' \in ON_{s'}(\sigma'(\alpha') = \sigma'(\beta') \to \alpha' = \beta')$. As an exercise, prove this.)

Theorem 5: $\forall \alpha' \in ON_{s'} \exists \alpha \in ON_s \; \alpha' = f(\alpha)$.

Proof: Case 1: $\alpha' = 0_{s'}$. Then $\alpha' = f(0_s)$. Case 2: $\alpha' = \omega_{s'}$. Then $\alpha' = f(\omega_s)$. Case 3: $\alpha' = \sigma'(\beta')$. As an inductive hypothesis, suppose β' satisfies the theorem. Then we can pick a β in ON_s such that $\beta' = f(\beta)$. But then $f(\sigma(\beta)) = \sigma'(f(\beta)) = \sigma'(\beta') = \alpha'$. So α' satisfies the theorem.

Theorem 6: $\forall \alpha, \beta \in ON_s(\alpha <_s \beta \leftrightarrow f(\alpha) <_{s'} f(\beta))$.

Proof: We shall first prove that $\alpha <_s \beta$ only if $f(\alpha) <_{s'} f(\beta)$. We must again consider three cases. Case 1: $\alpha = 0_s$. Suppose $0_s <_s \beta$. And suppose that either $f(\beta) = f(0_s)$ or $f(\beta) <_{s'} f(0_s)$. In the former case, Theorem 4 implies that $\beta = 0_s$ – which is absurd given that $0_s <_s \beta$. In the latter case, $f(\beta) <_{s'} 0_{s'}$ – which is also absurd. It follows that $0_s <_s \beta$ only if $f(0_s) <_{s'} f(\beta)$. Case 2: $\alpha = \omega_s$. Suppose β is the $<_s$-first member of ON_s such that $\omega_s <_s \beta$ and $f(\beta) <_{s'} f(\omega_s)$. Then $f(\beta) <_{s'} \omega_{s'}$. If $f(\beta) = 0_{s'}$, then Theorem 4 implies that $\beta = 0_s$ – which is absurd given that $\omega_s <_s \beta$. So we may pick a δ' such that $f(\beta) = \sigma'(\delta')$. And Theorem 5 allows us to pick a δ such that $\delta' = f(\delta)$. So $f(\beta) = \sigma'(f(\delta)) = f(\sigma(\delta))$. Then, by Theorem 4, $\beta = \sigma(\delta)$ and, hence, $\delta <_s \beta$. Furthermore, $f(\delta) <_{s'} f(\beta)$ and, hence, $f(\delta) <_{s'} f(\omega_s)$. So, by the minimality of β, either $\omega_s = \delta$ or $\delta <_s \omega_s$. If the former, then $f(\omega_s) <_{s'} f(\omega_s)$ – which is absurd. If the latter, then $\delta <_s \omega_s <_s \sigma(\delta)$ – which is also absurd. So it turns out to be false that $\omega_s <_s \beta$ and $f(\beta) <_{s'} f(\omega_s)$. Furthermore, by Theorem 4, it is impossible to have both $\omega_s <_s \beta$ and $f(\beta) = f(\omega_s)$. So $\omega_s <_s \beta$ only if $f(\omega_s) <_{s'} f(\beta)$. Case 3: $\alpha = \sigma(\gamma)$. Suppose $\sigma(\gamma) <_s \beta$. Then $\gamma <_s \beta$ (since $\gamma <_s \sigma(\gamma)$). As an inductive hypothesis, suppose $f(\gamma) <_{s'} f(\beta)$. And suppose that either $f(\beta) = f(\sigma(\gamma))$ or $f(\beta) <_{s'} f(\sigma(\gamma))$. In the former case, Theorem 4 implies that $\beta = \sigma(\gamma)$ – which is absurd given that $\sigma(\gamma) <_s \beta$. In the latter case, since $f(\sigma(\gamma)) = \sigma'(f(\gamma))$, $f(\gamma) <_{s'} f(\beta) <_{s'} \sigma'(f(\gamma))$ – which is also absurd. It follows that $\sigma(\gamma) <_s \beta$ only if $f(\sigma(\gamma)) <_{s'} f(\beta)$. We have shown that, in each of the three possible cases, $f(\alpha) <_{s'} f(\beta)$ whenever $\alpha <_s \beta$. Since the proof of the converse is essentially the same as the above, I leave it as an exercise.

Theorems 4, 5, and 6 establish that f is an isomorphism between ON_s (ordered by $<_s$) and $ON_{s'}$ (ordered by $<_{s'}$). We must now deal with $\mathcal{D}_s$ and $\mathcal{D}_{s'}$.

Let the function Γ be defined recursively as follows (where $\alpha \in ON_s$, $x \in \mathcal{D}_s$, and $x' \in \mathcal{D}_{s'}$):

$$\Gamma(\alpha) = \{<x,x'> : F_s x \alpha \wedge F_{s'} x' f(\alpha) \wedge \forall y \in \mathcal{D}_s (y \in_s x \leftrightarrow \\ \exists \gamma <_s \alpha \exists y' \in \mathcal{D}_{s'} (<y,y'> \in \Gamma(\gamma) \wedge y' \in_{s'} x')) \wedge \forall y' \in \mathcal{D}_{s'} (y' \in_{s'} x' \leftrightarrow \\ \exists \gamma <_s \alpha \exists y \in \mathcal{D}_s (<y,y'> \in \Gamma(\gamma) \wedge y \in_s x))\}.$$

And let the binary relation G be defined as follows:

$$\forall x \in \mathcal{D}_s \forall x' \in \mathcal{D}_{s'} (Gxx' \leftrightarrow \exists \alpha \in ON_s\ <x,x'> \in \Gamma(\alpha)).$$

Theorem 7: $\forall \alpha \in ON_s \forall x \in \mathcal{D}_s \forall x',z' \in \mathcal{D}_{s'} ((<x,x'> \in \Gamma(\alpha) \wedge <x,z'> \in \Gamma(\alpha)) \rightarrow x' = z')$.

Proof: Suppose $<x,x'> \in \Gamma(\alpha)$ and $<x,z'> \in \Gamma(\alpha)$. Then $\forall y' \in \mathcal{D}_{s'} (y' \in_{s'} x' \leftrightarrow \exists \gamma <_s \alpha \exists y \in \mathcal{D}_s (<y,y'> \in \Gamma(\gamma) \wedge y \in_s x))$ and $\forall y' \in \mathcal{D}_{s'} (y' \in_{s'} z' \leftrightarrow \exists \gamma <_s \alpha \exists y \in \mathcal{D}_s (<y,y'> \in \Gamma(\gamma) \wedge y \in_s x))$. So $\forall y' \in \mathcal{D}_{s'} (y' \in_{s'} x' \leftrightarrow y' \in_{s'} z')$. Since S' satisfies Axiom 2.4, this implies that $x' = z'$.

Theorem 8: $\forall \alpha, \beta \in ON_s \forall x \in \mathcal{D}_s \forall x',z' \in \mathcal{D}_{s'} ((<x,x'> \in \Gamma(\alpha) \wedge <x,z'> \in \Gamma(\beta)) \rightarrow \alpha = \beta)$.

Proof: Suppose $<x,x'> \in \Gamma(\alpha)$ and $<x,z'> \in \Gamma(\beta)$. Then $F_s x\alpha$ and $F_s x\beta$. Since S satisfies Theorem 2 of chapter 8, it follows that $\alpha=\beta$.

Theorem 9: $\forall x \in \mathfrak{D}_s \forall x',z' \in \mathfrak{D}_{s'}((Gxx' \wedge Gxz') \rightarrow x'=z')$.

Proof: Suppose Gxx' and Gxz'. Then we can pick an α and β such that $<x,x'> \in \Gamma(\alpha)$ and $<x,z'> \in \Gamma(\beta)$. But then, by Theorem 8, $\alpha=\beta$ and, hence, $<x,z'> \in \Gamma(\alpha)$. So, by Theorem 7, $x'=z'$.

Theorem 10: $\forall \alpha \in ON_s \forall x,z \in \mathfrak{D}_s \forall x' \in \mathfrak{D}_{s'}((<x,x'> \in \Gamma(\alpha) \wedge <z,x'> \in \Gamma(\alpha)) \rightarrow x=z)$.

Proof: I leave this as an exercise. (Hint: cf. the proof of Theorem 7.)

Theorem 11: $\forall \alpha,\beta \in ON_s \forall x,z \in \mathfrak{D}_s \forall x' \in \mathfrak{D}_{s'}((<x,x'> \in \Gamma(\alpha) \wedge <z,x'> \in \Gamma(\beta)) \rightarrow \alpha=\beta)$.

Proof: I leave this as an exercise. (Hint: cf. the proof of Theorem 8.)

Theorem 12: $\forall x,z \in \mathfrak{D}_s \forall x' \in \mathfrak{D}_{s'}((Gxx' \wedge Gzx') \rightarrow x=z)$.

Proof: I leave this as an exercise. (Hint: cf. the proof of Theorem 9.)

Theorem 13: $\forall x \in \mathfrak{D}_s \forall \alpha \in ON_s(F_s x\alpha \rightarrow \exists x' \in \mathfrak{D}_{s'} <x,x'> \in \Gamma(\alpha))$.

Proof: Suppose $F_s x\alpha$. Note that $\forall y' \in \mathfrak{D}_{s'} \forall \gamma \in ON_s(\exists y \in \mathfrak{D}_s <y,y'> \in \Gamma(\gamma) \rightarrow F_{s'} y' f(\gamma))$. So, by Theorem 6, $\forall y' \in \mathfrak{D}_{s'}(\exists \gamma <_s \alpha \exists y \in \mathfrak{D}_s <y,y'> \in \Gamma(\gamma) \rightarrow \exists \beta' <_{s'} f(\alpha) F_{s'} y'\beta')$. Furthermore, since S' satisfies Axiom 2.3, $\exists x' \in \mathfrak{D}_{s'} \forall y' \in \mathfrak{D}_{s'}(y' \in_{s'} x' \leftrightarrow (\exists \gamma <_s \alpha \exists y \in \mathfrak{D}_s(<y,y'> \in \Gamma(\gamma) \wedge y \in_s x) \wedge \exists \beta' <_{s'} f(\alpha) F_{s'} y'\beta'))$. (Since S' satisfies Axiom 2.3, the following is guaranteed: given any members of $\mathfrak{D}_{s'}$ to each of which $F_{s'}$ assigns a $\beta'<_{s'} f(\alpha)$, there is an $x' \in \mathfrak{D}_{s'}$ such that all and only the aforementioned members of $\mathfrak{D}_{s'}$ stand in $\in_{s'}$ to x'.) So $\exists x' \in \mathfrak{D}_{s'} \forall y' \in \mathfrak{D}_{s'}(y' \in_{s'} x' \leftrightarrow \exists \gamma <_s \alpha \exists y \in \mathfrak{D}_s (<y,y'> \in \Gamma(\gamma) \wedge y \in_s x))$. Pick such an x'. As an inductive hypothesis, suppose all the $<_s$-predecessors of α satisfy the theorem. And note that $\forall y \in \mathfrak{D}_s(y \in_s x \rightarrow \exists \gamma <_s \alpha F_s y\gamma)$ – since S satisfies Axiom 2.2. So, by our inductive hypothesis, $\forall y \in \mathfrak{D}_s(y \in_s x \rightarrow \exists \gamma <_s \alpha \exists y' \in \mathfrak{D}_{s'} <y,y'> \in \Gamma(\gamma))$. But then $\forall y \in \mathfrak{D}_s(y \in_s x \rightarrow \exists \gamma <_s \alpha \exists y' \in \mathfrak{D}_{s'}(<y,y'> \in \Gamma(\gamma) \wedge y' \in_{s'} x'))$. We must now prove the converse. Suppose $\exists \gamma <_s \alpha \exists y' \in \mathfrak{D}_{s'}(<y,y'> \in \Gamma(\gamma) \wedge y' \in_{s'} x')$. Then $\exists \gamma <_s \alpha \exists y' \in \mathfrak{D}_{s'}(<y,y'> \in \Gamma(\gamma) \wedge \exists \delta <_s \alpha \exists z \in \mathfrak{D}_s(<z,y'> \in \Gamma(\delta) \wedge z \in_s x))$. So $\exists y' \in \mathfrak{D}_{s'}(Gyy' \wedge \exists z \in \mathfrak{D}_s(Gzy' \wedge z \in_s x))$. It follows, by Theorem 12, that $y \in_s x$. We conclude that $\forall y \in \mathfrak{D}_s(y \in_s x \leftrightarrow \exists \gamma <_s \alpha \exists y' \in \mathfrak{D}_{s'}(<y,y'> \in \Gamma(\gamma) \wedge y' \in_{s'} x'))$. To complete the proof, we must show that $F_{s'} x' f(\alpha)$. We have already seen that $\forall y' \in \mathfrak{D}_{s'}(\exists \gamma <_s \alpha \exists y \in \mathfrak{D}_s <y,y'> \in \Gamma(\gamma) \rightarrow \exists \beta' <_{s'} f(\alpha) F_{s'} y'\beta')$. Furthermore, $\forall y' \in_{s'} x' \exists \gamma <_s \alpha \exists y \in \mathfrak{D}_s <y,y'> \in \Gamma(\gamma)$. It follows that $\forall y' \in_{s'} x' \exists \beta' <_{s'} f(\alpha)$

APPENDIX I   159

$F_{S'}y'\beta'$. Now suppose $f(\beta) <_{S'} f(\alpha)$. Then by Theorem 6, $\beta <_S \alpha$. So, since S satisfies Axiom 2.1, we may pick a $y \in_S x$ such that $\forall \gamma <_S \beta \neg F_S y\gamma$. But, as we have already seen, $\forall y \in \mathfrak{D}_S(y \in_S x \rightarrow \exists \gamma <_S \alpha F_S y\gamma)$. So we may pick a $\gamma <_S \alpha$ such that $F_S y\gamma$ and either $\beta <_S \gamma$ or $\beta = \gamma$. By our inductive hypothesis, we may pick a $y' \in \mathfrak{D}_{S'}$ such that $<y,y'> \in \Gamma(\gamma)$. Then $F_{S'}y'f(\gamma)$ and $y' \in_{S'} x'$. Furthermore, by Theorem 6, either $f(\beta) <_{S'} f(\gamma)$ or $f(\beta) = f(\gamma)$. Since S' satisfies Theorem 2 of chapter 8, $F_{S'}$ is a function. So $\forall \gamma' <_{S'} f(\beta) \neg F_{S'}y'\gamma'$. In more general terms, $\forall \beta' <_{S'} f(\alpha) \exists y' \in_{S'} x' \forall \gamma' <_{S'} \beta' \neg F_{S'}y'\gamma'$. Since S' satisfies Axiom 2.1, it follows that $F_{S'}x'f(\alpha)$. We conclude that $<x,x'> \in \Gamma(\alpha)$.

Theorem 14: $\forall x \in \mathfrak{D}_S \exists x' \in \mathfrak{D}_{S'} Gxx'$.

Proof: Pick an $x \in \mathfrak{D}_S$. Since S satisfies Theorem 3 of chapter 8, we may pick an $\alpha \in ON_S$ such that $F_S x\alpha$. Then, by Theorem 13, $\exists x' \in \mathfrak{D}_{S'}$ $<x,x'> \in \Gamma(\alpha)$. So $\exists x' \in \mathfrak{D}_{S'} Gxx'$.

Theorem 15: $\forall x' \in \mathfrak{D}_{S'} \forall \alpha \in ON_S(F_{S'}x'f(\alpha) \rightarrow \exists x \in \mathfrak{D}_S <x,x'> \in \Gamma(\alpha))$.

Proof: I leave this as an exercise. (Hint: cf. the proof of Theorem 13.)

Theorem 16: $\forall x' \in \mathfrak{D}_{S'} \exists x \in \mathfrak{D}_S Gxx'$.

Proof: I leave this as an exercise. (Hint: cf. the proof of Theorem 14.)

Theorems 9 and 14 allow us to define the function g as follows: $\forall x \in \mathfrak{D}_S \forall x' \in \mathfrak{D}_{S'}(g(x) = x' \leftrightarrow Gxx')$. Theorems 12 and 16 establish that g is a pairing between the members of $\mathfrak{D}_S$ and the members of $\mathfrak{D}_{S'}$. We must now show that g preserves relative structural position.

Theorem 17: $\forall x,y \in \mathfrak{D}_S(y \in_S x \leftrightarrow g(y) \in_{S'} g(x))$.

Proof: Pick an $x \in \mathfrak{D}_S$. Since S satisfies Theorem 3 of chapter 8, we may pick an $\alpha \in ON_S$ such that $F_S x\alpha$. Then $\forall y' \in \mathfrak{D}_{S'}(y' \in_{S'} g(x) \leftrightarrow \exists \gamma <_S \alpha \exists y \in \mathfrak{D}_S(<y,y'> \in \Gamma(\gamma) \land y \in_S x))$. Suppose $g(y) \in_{S'} g(x)$. Then $\exists \gamma <_S \alpha \exists z \in \mathfrak{D}_S (<z,g(y)> \in \Gamma(\gamma) \land z \in_S x)$. So $\exists z \in \mathfrak{D}_S(g(z) = g(y) \land z \in_S x)$. It follows, by Theorem 12, that $y \in_S x$. We have shown that $g(y) \in_{S'} g(x)$ only if $y \in_S x$. To prove the converse, suppose $y \in_S x$. Then $\exists \gamma <_S \alpha \exists y' \in \mathfrak{D}_{S'}(<y,y'> \in \Gamma(\gamma) \land y' \in_{S'} g(x))$. So $\exists y' \in \mathfrak{D}_{S'}(y' = g(y) \land y' \in_{S'} g(x))$ and, hence, $g(y) \in_{S'} g(x)$. We conclude that $y \in_S x$ if and only if $g(y) \in_{S'} g(x)$.

Theorem 18: $\forall x \in \mathfrak{D}_S \forall \alpha \in ON_S(F_S x\alpha \leftrightarrow F_{S'} g(x) f(\alpha))$.

Proof: It follows from the definition of Γ that $F_S x\alpha$ only if $F_{S'} g(x) f(\alpha)$. To prove the converse, suppose $F_{S'} g(x) f(\alpha)$. Since S satisfies Theorem 3 of chapter 8, we may pick a $\beta \in ON_S$ such that $F_S x\beta$. Then $F_{S'} g(x) f(\beta)$. So, since $F_{S'}$ satisfies Theorem 2 of chapter 8, $f(\alpha) = f(\beta)$. But then, by Theorem 4, $\alpha = \beta$. So $F_S x\alpha$.

160   APPENDIX I

We have shown that any two models of T_{1*} will be isomorphic to one another. So T_{1*} characterizes exactly *one* abstract structure and, hence, given any proposition φ in the language of T_{1*}, φ will be true either in all models of T_{1*} or in none. In particular, CH will be true either in all models of T_{1*} or in none. So either CH or $\neg$CH is a logical consequence of the axioms of T_{1*}.

APPENDIX II

We can use some techniques of James Van Aken to reformulate our theory T_2' in an elegant way. $\Re(x)$ will be the rank of x (i.e., the ordinal which indicates the position of x in our iterative hierarchy). We shall use '$x \in ON$' to abbreviate '$\exists y \; x = \Re(y)$'.

Axiom I: $\forall X (\exists y \forall x (Xx \rightarrow \Re(x) \in \Re(y)) \rightarrow \exists y \forall x (x \in y \leftrightarrow Xx))$.

Axiom II: $\forall x,y (\Re(x) \subset \Re(y) \leftrightarrow \exists z \in y \forall w \in x \; \Re(w) \in \Re(z))$.

Axiom III: $\forall x \in ON \forall y \in x \; y \in ON$.

Axiom IV: $\forall x,y (\forall z (z \in x \leftrightarrow z \in y) \rightarrow x=y)$.

Axiom V: $\exists x \in ON (\exists y \; y \in x \land \forall y \in x \exists y' \in x \; y \in y')$.

Axiom VI: $\forall x \in ON \exists y \in ON \; x \in y$.

Axiom VII: $\forall X (\forall y \exists ! y' \; X<y,y'> \rightarrow \forall x \exists z \in ON \forall w \in ON (\exists y \in x \; X<y,w> \rightarrow w \in z))$.

It will turn out that our ranks are none other than the Mirimanoff ordinals. We shall use '$x \in ORD$' to express that x is such an ordinal (i.e., that x is internally well-ordered by $\in$ and that every member of x is also a subset of x).

Theorem 1: $\forall X (\forall x (\forall y \in x \; Xy \rightarrow Xx) \rightarrow \forall x \; Xx)$.

Proof: Suppose $\forall x (\forall y \in x \; Xy \rightarrow Xx)$ but $\neg Xs$. It follows that $\forall x (\forall y (\neg Xy \rightarrow \Re(x) \in \Re(y)) \rightarrow \Re(x) \in \Re(s))$ and, *a fortiori*, that $\forall x ((x \notin x \land \forall y (\neg Xy \rightarrow \Re(x) \in \Re(y))) \rightarrow \Re(x) \in \Re(s))$. So Axiom I allows us to pick a u such that $\forall x (x \in u \leftrightarrow (x \notin x \land \forall y (\neg Xy \rightarrow \Re(x) \in \Re(y))))$. Since $\neg Xs$, our initial assumption allows us to pick a $t \in s$ such that $\neg Xt$. Then $\forall w \in u \; \Re(w) \in \Re(t)$. More generally, $\forall y (\neg Xy \rightarrow \exists z \in y \forall w \in u \; \Re(w) \in \Re(z))$. But then, by Axiom II, $\forall y (\neg Xy \rightarrow \Re(u) \in \Re(y))$. Since this implies the absurd result that $u \in u$ if and only if $u \notin u$, we may conclude that (contrary to our second assumption) Xs and, more generally, that $\forall x Xx$.

Theorem 2: $\forall y \forall x \in y \; \Re(x) \in \Re(y)$.

Proof: Suppose every member of y satisfies the theorem – that is, $\forall y' \in y \forall w \in y' \; \Re(w) \in \Re(y')$. We want to show that y also satisfies the theorem. Accordingly, suppose $x \in y$. Then $\forall w \in x \; \Re(w) \in \Re(x)$. So $\exists z \in y \forall w \in x \; \Re(w) \in \Re(z)$. Hence, by Axiom II, $\Re(x) \in \Re(y)$ and, more generally, $\forall x \in y \; \Re(x) \in \Re(y)$. Now apply Theorem 1.

Theorem 3: $\forall x \; \Re(x) \notin \Re(x)$.

Proof: By Axiom II, $\forall x (\Re(x) \in \Re(x) \leftrightarrow \exists y \in x \forall w \in x \; \Re(w) \in \Re(y))$. So $\forall x (\Re(x) \in \Re(x) \rightarrow \exists y \in x \; \Re(y) \in \Re(y))$ – which is equivalent to $\forall x (\forall y \in x \; \Re(y) \notin \Re(y) \rightarrow \Re(x) \notin \Re(x))$. Now apply Theorem 1.

Theorem 4: $\forall x,y,z((\Re(x) \in \Re(y) \land \Re(y) \in \Re(z)) \rightarrow \Re(x) \in \Re(z))$.

Proof: Suppose every member of x satisfies the theorem – that is, $\forall x' \in x \forall y',z'((\Re(x') \in \Re(y') \land \Re(y') \in \Re(z')) \rightarrow \Re(x') \in \Re(z'))$. We want to show that x also satisfies the theorem. Accordingly, suppose $(\Re(x) \in \Re(y) \land \Re(y) \in \Re(z))$. Then, by Axiom II, $\exists y' \in y \forall x' \in x \; \Re(x') \in \Re(y')$ and $\exists z' \in z \forall y' \in y \; \Re(y') \in \Re(z')$. It follows that $\exists z' \in z \exists y' \in y \forall x' \in x (\Re(x') \in \Re(y') \land \Re(y') \in \Re(z'))$. But then, by our first assumption, $\exists z' \in z \forall x' \in x \; \Re(x') \in \Re(z')$. So, by Axiom II, $\Re(x) \in \Re(z)$. Now apply Theorem 1.

Theorem 5: $\forall x \in \mathrm{ON} \forall y \in x \; y \subset x$.

Proof: Apply Axiom III and Theorem 4.

Theorem 6: $\forall X (\exists x \in \mathrm{ON} \; Xx \rightarrow \exists x \in \mathrm{ON}(Xx \land \forall y \in x \; \neg Xy))$.

Proof: Suppose that $\exists x \in \mathrm{ON} \; Xx$. Theorem 1 implies that $(\forall x (\forall y \in x \; (y \in \mathrm{ON} \rightarrow \neg Xy) \rightarrow (x \in \mathrm{ON} \rightarrow \neg Xx)) \rightarrow \forall x \in \mathrm{ON} \; \neg Xx)$. So $\neg \forall x (\forall y \in x \; (y \in \mathrm{ON} \rightarrow \neg Xy) \rightarrow (x \in \mathrm{ON} \rightarrow \neg Xx))$ – i.e., $\exists x \in \mathrm{ON}(Xx \land \forall y \in x(y \in \mathrm{ON} \rightarrow \neg Xy))$. Axiom III then implies that $\exists x \in \mathrm{ON}(Xx \land \forall y \in x \; \neg Xy)$.

Theorem 7: $\forall x,y \in \mathrm{ON}(x \in y \lor y \in x \lor x=y)$.

Proof: Suppose the theorem is false. By Theorem 6, pick a minimal $x \in \mathrm{ON}$ such that $\neg \forall y \in \mathrm{ON}(x \in y \lor y \in x \lor x=y)$. (A rank is minimal with respect to some property if it has the property, but all its predecessors lack it.) Apply Theorem 6 again to pick a minimal $y \in \mathrm{ON}$ such that $\neg (x \in y \lor y \in x \lor x=y)$. Suppose $x' \in x$. Then $x' \neq y$ and, by Theorem 5, $y \notin x'$. So, by Axiom III and the minimality of x, $x' \in y$. Suppose $y' \in y$. Then $y' \neq x$ and, by Theorem 5, $y \notin x'$. So, by Axiom III and the minimality of y, $y' \in x$. We have shown that x and y have exactly the same members. So, by Axiom IV, $x=y$ – contradicting our choice of x and y. It follows that the theorem is not false after all.

Theorem 8: $\forall x \in \text{ON} \ x \in \text{ORD}$.

Proof: Theorems 3, 4, 6, and 7 say that $\in$ forms a well-ordering between our ranks. So Axiom III implies that our ranks are internally well-ordered by $\in$. Furthermore, Theorem 5 says that every member of a rank is a subset of that rank. So our ranks are all Mirimanoff ordinals.

Theorem 9: $\forall x,y (\mathcal{R}(x) \in \mathcal{R}(y) \rightarrow \exists z \in y \ \mathcal{R}(z) \notin \mathcal{R}(x))$.

Proof: Suppose the theorem is false. By Theorem 6, pick the first $\mathcal{R}(y)$ such that $\exists x (\mathcal{R}(x) \in \mathcal{R}(y) \land \forall z \in y \ \mathcal{R}(z) \in \mathcal{R}(x))$. Pick such an x. Then, by Axiom II, $\exists z \in y \forall w \in x \ \mathcal{R}(w) \in \mathcal{R}(z)$. So $\exists z (\mathcal{R}(z) \subset \mathcal{R}(x) \land \forall w \in x \ \mathcal{R}(w) \in \mathcal{R}(z))$ – contradicting the minimality of $\mathcal{R}(y)$.

Theorem 10: $\forall x,y ((\forall z \in y \ \mathcal{R}(z) \in \mathcal{R}(x) \land \forall z \in \mathcal{R}(x) \exists w \in y \ \mathcal{R}(w) \notin z) \rightarrow \mathcal{R}(x) = \mathcal{R}(y))$.

Proof: Suppose $\forall z \in y \ \mathcal{R}(z) \in \mathcal{R}(x)$ and $\forall z \in \mathcal{R}(x) \exists w \in y \ \mathcal{R}(w) \notin z$. And suppose that $\mathcal{R}(x) \in \mathcal{R}(y)$. Then, by Theorem 9, $\exists z \in y \ \mathcal{R}(z) \notin \mathcal{R}(x)$ – contrary to our first assumption. So $\mathcal{R}(x) \notin \mathcal{R}(y)$. Now assume that $\mathcal{R}(y) \in \mathcal{R}(x)$. Then our second assumption implies that $\exists w \in y \ \mathcal{R}(w) \notin \mathcal{R}(y)$ – contrary to Theorem 2. So $\mathcal{R}(y) \notin \mathcal{R}(x)$. It follows by Theorem 7, that $\mathcal{R}(x) = \mathcal{R}(y)$.

Theorem 11: $\forall x \ \mathcal{R}(x) = \mathcal{R}(\mathcal{R}(x))$.

Proof: Suppose $\forall z \in \mathcal{R}(x) \ z = \mathcal{R}(z)$. Then, by Theorem 3, $\forall z \in \mathcal{R}(x) \exists w \in \mathcal{R}(x) \ \mathcal{R}(w) \notin z$. Furthermore, $\forall z \in \mathcal{R}(x) \ \mathcal{R}(z) \in \mathcal{R}(x)$. Hence, by Theorem 10, $\mathcal{R}(x) = \mathcal{R}(\mathcal{R}(x))$. So if we assume that the theorem is false and if, by Theorem 6, we let $\mathcal{R}(x)$ be the first rank which fails to satisfy the theorem, then $\mathcal{R}(x)$ will turn out to satisfy the theorem after all. It follows that the theorem is true.

Theorem 12: $\forall x (\forall y \in x (y \subset x \land y \in \text{ON}) \rightarrow x \in \text{ON})$.

Proof: Suppose $\forall y \in x (y \subset x \land y \in \text{ON})$. By Theorem 11, $\forall y \in x \ y = \mathcal{R}(y)$. So, by Theorem 2, $\forall y \in x \ y \in \mathcal{R}(x)$. Further, by Axiom III and Theorem 9, $\forall y \in \mathcal{R}(x) \exists z \in x \ \mathcal{R}(z) \notin y$. So, by Axiom III and Theorem 7, $\forall y \in \mathcal{R}(x) \exists z \in x \ (y \in \mathcal{R}(z) \lor y = \mathcal{R}(z))$. This implies, by Theorem 11, that $\forall y \in \mathcal{R}(x) \exists z \in x (y \in z \lor y = z)$. If $y \in z$ and $z \in x$, then $y \in x$. So $\forall y \in \mathcal{R}(x) \ y \in x$. We have shown that x and $\mathcal{R}(x)$ have exactly the same members. Hence, by Axiom IV, $x = \mathcal{R}(x)$.

Theorem 13: $\forall x \in \text{ORD} \forall y \in x \ y \in \text{ORD}$.

Proof: Suppose $x \in \text{ORD}$ and $y \in x$. Then $y \subset x$ and, since x is well-ordered by $\in$, so is y. Now suppose $w \in z$ and $z \in y$. Then $z \in x$ and, hence, $w \in x$. But then $w \in y$ since $\in$ well orders x. So $z \subset y$ and, more generally, $\forall z \in y \ z \subset y$. Hence $y \in \text{ORD}$.

Theorem 14: $\forall X(\exists x \in \text{ORD} \ Xx \to \exists x \in \text{ORD}(Xx \land \forall y \in x \ \neg Xy))$.

Proof: We can copy our proof of Theorem 6 almost word for word. We need only replace 'ON' with 'ORD' and replace the reference to Axiom III with a reference to Theorem 13.

Theorem 15: $\forall x \in \text{ORD} \ x \in \text{ON}$.

Proof: Suppose $\exists x \in \text{ORD} \ x \notin \text{ON}$. Then, by Theorem 14, $\exists x \in \text{ORD} \ (x \notin \text{ON} \land \forall y \in x \ y \in \text{ON})$. But this implies that $\exists x(\forall y \in x(y \subset x \land y \in \text{ON}) \land x \notin \text{ON})$ – contrary to Theorem 12. We may conclude that $\neg \exists x \in \text{ORD} \ x \notin \text{ON}$ – that is, $\forall x \in \text{ORD} \ x \in \text{ON}$.

Note that we have proved all of the above theorems without the aid of Axioms V, VI, and VII. This means that our identification of ON and ORD does not depend on any assumptions about the "height" of our iterative hierarchy. (The study of set theories which are free of such assumptions was pioneered by Richard Montague and Dana Scott.[114])

NOTES

1. A. D. Aleksandrov et al., eds., *Mathematics: Its Content, Methods, and Meaning*, vol. 1, p. 63.
2. Kenneth Kunen, *Set Theory: An Introduction to Independence Proofs*, p. xi. John Mayberry goes even farther ("What Are Numbers?" p. 353): "... set theory is not really, or not just, a foundation for mathematics. It simply *is* modern mathematics."
3. Robert G. Bartle, *The Elements of Real Analysis*, p. 1.
4. Paul J. Cohen, *Set Theory and the Continuum Hypothesis*, p. 50.
5. Aleksandrov et al., vol. 3, pp. 5-6.
6. Philip J. Davis and Reuben Hersh, *The Mathematical Experience*, pp. 29-30.
7. Stanislaw M. Ulam, *Adventures of a Mathematician*, pp. 288-289. Cited in Davis and Hersh, p. 21.
8. Ulam, p. 288.
9. Aleksandrov et al., vol. 1, p. 63.
10. Hermann Weyl, *Philosophy of Mathematics and Natural Science*, pp. 8-13.
11. Bodo Pareigis, *Categories and Functors*, p. vii.
12. Cf. Abraham A. Fraenkel et al., *Foundations of Set Theory*, pp. 143-145.
13. Philip Kitcher, *The Nature of Mathematical Knowledge*, pp. 4-5.
14. Imre Lakatos, *Proof and Refutations: The Logic of Mathematical Discovery*, p. 2.
15. William Chase Greene, ed., *Scholia Platonica*, p. 115.
16. Edward Hiller, ed., *Theonis Smyrnaei: Expositio rerum mathematicarum ad legendum Platonem utilium*, p. 19.
17. Cf. Jacob Klein's magnificent *Greek Mathematical Thought and the Origin of Algebra*.
18. Edith Dudley Sylla, "The Oxford Calculators," p. 563.
19. Klein, chap. 11.
20. Ibid., pp. 171-176.
21. Carl B. Boyer, *The History of the Calculus and its Conceptual Development*, p. 190.
22. Leroy E. Loemker, ed., *Gottfried Wilhelm Leibniz: Philosophical Papers and Letters*, pp. 28-31.

23. Boyer, p. 243.
24. Cf. A. P. Youschkevitch, "The Concept of Function up to the Middle of the 19th Century," pp. 75–76.
25. D. J. Struik, ed., *A Source Book in Mathematics, 1200–1800*, p. 361.
26. Charles H. Edwards, *The Historical Development of the Calculus*, p. 307.
27. Cf. Joseph W. Dauben's essay "The Trigonometric Background to Georg Cantor's Theory of Sets."
28. Ibid., pp. 194–197.
29. Michael Hallett, *Cantorian Set Theory and Limitation of Size*, p. 3.
30. Dauben, pp. 213–215.
31. Georg Cantor, *Contribution to the Founding of the Theory of Transfinite Numbers*, p. 86.
32. Hermann Weyl, *The Continuum*, p. 93.
33. Jean van Heijenoort, ed., *From Frege to Gödel: A Source Book in Mathematical Logic, 1879–1931*, p. 114. See also Richard Dedekind, *Essays on the Theory of Numbers*, p. 47.
34. Cf. Kitcher, *The Nature of Mathematical Knowledge*, pp. 267–268; Hallett, p. 244.
35. Cf. Gottlob Frege, *The Foundations of Arithmetic*, pp. 74–80; Bertrand Russell, *The Principles of Mathematics*, pp. 114–116. It would be wrong not to mention that Frege explicitly affirmed the existence of sets whose members are themselves sets. I have chosen not to emphasize this facet of his work because the influence of Fregean set theory on the details of current set theory is unclear. Cantor's vital contributions are much easier to make out. (Hence the practice of referring invidiously to Fregean sets as "logical" rather than "mathematical.") Frege's notion of set is roughly as follows. To every concept there corresponds a set whose members are those entities which fall under the concept. For example, to the concept "being human" there corresponds the set $\{x:x$ is human$\}$ whose members are those entities which fall under the concept "being human." (That is, $\{x:x$ is human$\}$ is the set of human beings.) On this view, the existence of sets of sets follows from the simple fact that some concepts have sets falling under them. For example, all two-membered sets fall under the concept "having two members." So the set $\{x:x$ has two members$\}$ corresponding to this concept certainly has sets as members. Since we are exploring the influence of mundane set conceptions on the development of mathematical set theory, two comments are pertinent here. First, the only formulations of Frege's doctrine which have any claim to being commonsensical happen also to be inconsistent and, hence, are of no immediate mathematical value. (Cf. the discussion of the Zermelo/Russell antinomy in §2 of chapter 6.) Second, Frege's official formulation of his doctrine can scarcely be considered a natural development of everyday notions. Rather, it represents a revolutionary break with a vision of logic and language which had dominated western thought since the time of Aristotle. Frege views concepts as *functions* which map objects onto truth val-

ues. For example, "being human" maps Socrates onto Truth and Lassie onto Falsity (since Socrates is human, but Lassie is not). The set {x:x is human} is really the *value range* of the function "being human" and, hence, is some sort of aggregation of pairs such as <Socrates, Truth> and <Lassie, Falsity>. The claim that Socrates is a member of {x:x is human} must be regarded as an abbreviation of the claim that the pair <Socrates, Truth> is part of the value range of "being human." The bizarreness of this doctrine cannot detract from its theoretical fertility. (Cf. Michael Dummett, *The Interpretation of Frege's Philosophy*, chap. 8.) My point is simply that Frege's theory cannot plausibly be considered a natural extension of everyday set talk. (So much the worse for everyday set talk!)

36. Russell, *The Principles of Mathematics*, p. 286.
37. Cf. Kitcher, *The Nature of Mathematical Knowledge*, pp. 259-268.
38. Cited in Max Black, "The Elusiveness of Sets," p. 615.
39. Ibid., p. 634.
40. Ibid., pp. 628-629.
41. Peter Simons, "Numbers and Manifolds," p. 165.
42. Black, p. 629.
43. Ibid., p. 631.
44. Ibid., p. 632.
45. For a more complete defense of this claim, see George Bealer, *Quality and Concept*, pp. 102-106. Cf. also Simons, "Numbers and Manifolds," pp. 187-195, and "Plural Reference and Set Theory," pp. 208-211; Ruth Barcan Marcus, "Classes, Collections, and Individuals," passim; Russell, *The Principles of Mathematics*, §§71 and 74.
46. Ernst Zermelo, "Investigations in the Foundations of Set Theory I."
47. Simons, "Numbers and Manifolds," p. 194.
48. Black, p. 633. For an attempt at amelioration, see Mayberry, "What Are Numbers?" pp. 342-343.
49. Simons, "Plural Reference and Set Theory," pp. 209-212.
50. Cf. George Bealer, "Foundations Without Sets"; Frederic B. Fitch, "Attribute and Class"; Nicolas D. Goodman, "A Genuinely Intensional Set Theory"; Stephen Pollard and Norman M. Martin, "Mathematics for Property Theorists"; Stephen Pollard, "Transfinite Recursion in a Theory of Properties" and "Identity Criteria"; Rolf Schock, "New Foundations for Concept Theory." The relevant work by Michael Jubien is currently unpublished.
51. Simons, "Plural Reference and Set Theory," p. 210. Cf. also Marcus, p. 227.
52. See, in particular, Alfred Tarski, "A General Method in Proofs of Undecidability." The words 'theory' and 'theorem' are used in a special sense in what follows. A *theory* is a collection of sentences which is closed under logical consequence. (If T is a theory and a sentence S follows logically from sentences in T, then S is also in T.) Every sentence in a theory T is said to be a *theorem* of T. When understood in this way, Zermelo set theory would consist of the axioms of Z (introduced in chapter 3) together with all the set

theoretic sentences which follow from those axioms. The theorems of Z would be the axioms of Z plus their logical consequences.

53. Cf. Giuseppe Peano, "Le definizioni per astrazione"; Weyl, *Philosophy of Mathematics and Natural Science*, pp. 8-13; Paul Lorenzen, *Formal Logic*, pp. 105-111 and "Equality and Abstraction." See also Ignacio Angelelli, "Abstraction, Looking-Around and Semantics" and Stephen Pollard, "What Is Abstraction?"

54. Cf. Gottlob Frege, "Review of Dr. E. Husserl's *Philosophy of Arithmetic*" and Edmund Husserl, *Logical Investigations*, Investigation II, chap. 3.

55. For a fuller discussion of the relation between set formation and abstraction, see Stephen Pollard, "Weyl on Sets and Abstraction."

56. Cf. Daniel A. Bonevac's *Reduction in the Abstract Sciences*.

57. Paul J. Cohen, "Comments on the Foundations of Set Theory," pp. 11 and 15.

58. Cf. Jonathan Lear, "Sets and Semantics," p. 88.

59. Cf. pp. ix-x of Dana Scott's Foreword to J. L. Bell, *Boolean-Valued Models and Independence Proofs in Set Theory*.

60. Edmund Husserl, *Formal and Transcendental Logic*, Part I, chap. 3 ("Theory of Deductive Systems and Theory of Multiplicities").

61. Kunen, *Set Theory: An Introduction to Independence Proofs*, p. xi.

62. Cf. Kurt Gödel, *The Consistency of the Continuum Hypothesis*.

63. Kunen, p. 171. See also Cohen, "Comments on the Foundations of Set Theory," p. 12; Fraenkel et al., pp. 108-109; and Penelope Maddy, "Believing the Axioms," pp. 492-495.

64. Cf. Cohen, *Set Theory and the Continuum Hypothesis*, chap. 4.

65. Cf. Kunen, pp. 232-235. For an excellent popular exposition of the proof (co-authored by Cohen himself), see Davis and Hersh, *The Mathematical Experience*, pp. 230-236. Mary Tiles offers a fine overview of the Gödel-Cohen proofs in chap. 8 of *The Philosophy of Set Theory*.

66. Cf. Fraenkel et al., pp. 104-105.

67. Abraham Robinson, "Formalism 64," pp. 232 and 233.

68. Cohen, "Comments on the Foundations of Set Theory," p. 13.

69. Cf. Georg Kreisel, "Observations on Popular Discussions of Foundations."

70. Cohen, "Comments on the Foundations of Set Theory," p. 12.

71. For a vitriolic discussion of these first two points, see Kreisel, "Observations on Popular Discussions of Foundations," pp. 194-195.

72. Reasons for thinking that GCH will eventually be considered true are given in Fraenkel et al., pp. 106-108. Reasons for thinking that CH will eventually be considered false are given in Kurt Gödel, "What Is Cantor's Continuum Problem?" pp. 478-480. (Gödel writes, "... one has good reason for suspecting that the role of the continuum problem in set theory will be to lead to the discovery of new axioms which will make it possible to disprove Cantor's conjecture.") Cf. also Cohen (!), *Set Theory and the Continuum Hypothesis*, p. 151. For a good recent overview of the debate, see Maddy, pp. 494-500, 746-748. See also Tiles, chap. 9.

NOTES TO PAGES 72-99   169

73. Cf. Michael Dummett, "The Philosophical Basis of Intuitionistic Logic."

74. Ibid. (See Dummett, *Truth and Other Enigmas*, pp. 217-218 or Paul Benacerraf and Hilary Putnam, eds., *Philosophy of Mathematics: Selected Readings*, pp. 99-100.)

75. Cf. Michael Dummett, *Elements of Intuitionism*, pp. 370-371 and *Truth and Other Enigmas*, p. 366. For a detailed application of nonclassical semantics to classical set theory, see Lear.

76. Cf. Robinson, p. 235.

77. Cf. Cohen, "Comments on the Foundations of Set Theory," p. 15.

78. Cf. Georg Kreisel and J. L. Krivine, *Elements of Mathematical Logic: Model Theory*, pp. 220-226.

79. Cf. Kreisel, "Two Notes on the Foundations of Set Theory," p. 107 and "Observations on Popular Discussions of Foundations," p. 197. For recent overviews of the accomplishments of Hilbertian formalism, see Wilfried Sieg, "Hilbert's Program Sixty Years Later" and Stephen G. Simpson, "Partial Realizations of Hilbert's Program." For an extensive philosophical reappraisal, see Michael Detlefsen, *Hilbert's Program*.

80. Bernhard Rang and Wolfgang Thomas, "Zermelo's Discovery of the 'Russell Paradox'."

81. Cf. Georg Cantor, "Letter to Dedekind" and Cesare Burali-Forti, "A Question on Transfinite Numbers."

82. Cf. Stephen Pollard, "A Peculiarity of the Empty Set."

83. Cf. Zermelo, "Investigations in the Foundations of Set Theory I."

84. Cf. ibid., pp. 204-205.

85. But see Kreisel and Krivine, pp. 171-174 and 214-220.

86. Cf. George Boolos, "To Be is to Be a Value of a Variable (or to Be Some Values of Some Variables)" and "Nominalist Platonism." For a critique of Boolos' approach see Michael D. Resnik, "Second-Order Logic Still Wild." I find Resnik's criticisms to be not at all telling. One problem is Resnik's assimilation of plurals and *predicates* (p. 78). It is much more illuminating (and much closer to Boolos' intentions) to compare plurals with *singular* expressions from the same grammatical category. Let me explain. Bound occurrences of singular pronouns range (minimally) over the referents of appropriate singularly referring expressions. A bound occurrence of 'he' could range over the referents of 'Charles Parsons', 'the chief justice of the U.S. Supreme Court', and so on. Bound occurrences of plural pronouns range (minimally) over the referents of appropriate plurally referring expressions. A bound occurrence of 'they' could range over the referents of 'Charles Parsons and Michael Resnik', 'the justices of the U.S. Supreme Court', and so on. The sort of things to which singular quantifications commit us are just the sort of things to which singularly referring expressions refer. The sort of things to which plural quantifications commit us are just the sort of things to which plurally referring expressions refer. The crucial question then is: To what do plurally referring expressions refer? I should have thought the answer is uncontroversial: the only (referential) difference between singular and plural ex-

pressions is that the former refer to things one at a time, while the latter refer to things more than one at a time. 'Charles Parsons' refers to Charles Parsons. 'Charles Parsons and Michael Resnik' refers simultaneously to Charles Parsons and Michael Resnik. So the sort of things to which plural quantifiers commit us are exactly the sort of things to which singular quantifiers commit us. The sentence 'Some philosophers sang a duet last night' commits me to objects of the same sort as the referents of expressions which can reasonably be substituted for 'Some philosophers'. So this sentence commits me to objects such as Charles Parsons and Michael Resnik (as in 'Charles Parsons and Michael Resnik sang a duet last night'). I do not think any "fledgling theory of divided reference" (to use Resnik's phrase) is being invoked here. It is just a matter of reminding ourselves how plurals (*not predicates*) operate in English.

87. Cf. Dummett, *Truth and Other Enigmas*, chap. 10.

88. An even stronger result is available. See Stephen Pollard, "Plural Quantification and the Axiom of Choice," p. 397.

89. For an excellent discussion of second order languages from the point of view of the standard semantics see Stewart Shapiro, "Second-Order Languages and Mathematical Practice." Shapiro argues powerfully that "... an adequate formalization of such branches of mathematics as arithmetic, real analysis, and set theory must involve (at least) a second order language." And he concludes that "... the natural underlying *logic* of these branches is (at least) second order." For an intelligent critique of this position, see Steven J. Wagner, "The Rationalist Conception of Logic," §5. For a lively discussion of second order set theories, see Imre Lakatos, ed., *Problems in the Philosophy of Mathematics,* pp. 82-117 and 138-152.

90. Cf. Charles Parsons, "What is the Iterative Conception of Set?" Parsons' essay is, in part, a response to Hao Wang's paper "The Concept of Set."

91. Cf. Dimitry Mirimanoff, "Les antinomies de Russell et de Burali-Forti et le problème fondamental de la théorie des ensembles." See also Hallett, pp. 185-197.

92. Kazimierz Kuratowski, "Sur la notion d'ordre dans la théorie des ensembles."

93. Cf. Shapiro, p. 741.

94. Boolos, "The Iterative Conception of Set." See also Stephen Pollard, "Plural Quantification and the Iterative Concept of Set."

95. Dana S. Scott, "Axiomatizing Set Theory."

96. James Van Aken, "Axioms for the Set-Theoretic Hierarchy."

97. Henri Poincaré, *Science and Hypothesis*, p. 20.

98. Russell, *Introduction to Mathematical Philosophy*, p. 59.

99. Paul Benacerraf, "What Numbers Could Not Be."

100. Philip Kitcher, "The Plight of the Platonist."

101. Benacerraf, pp. 69-70. (Or see Paul Benacerraf and Hilary Putnam, eds., *Philosophy of Mathematics: Selected Readings,* pp. 290-291.)

102. Dedekind, *Essays on the Theory of Numbers*, p. 68.

103. Michael D. Resnik, "Mathematics as a Science of Patterns: Ontology and Reference," p. 530. An excellent discussion of mathematical structuralism is also to be found in Charles Parsons' essay "The Structuralist View of Mathematical Objects." For Resnik's most recent views, see his "Mathematics from the Structural Point of View."

104. Michael D. Resnik, "Mathematics as a Science of Patterns: Ontology and Reference," p. 539.

105. Michael D. Resnik, "Mathematical Knowledge and Pattern Cognition," p. 38.

106. Resnik, "Mathematics as a Science of Patterns: Epistemology."

107. Resnik, "Mathematical Knowledge and Pattern Cognition," p. 29.

108. Resnik, "Mathematics as a Science of Patterns: Ontology and Reference," p. 529.

109. Resnik, "Mathematical Knowledge and Pattern Cognition," p. 34.

110. Ibid., pp. 29, 35.

111. I should explain how interpretation functions are to handle second order quantifiers. If the interpretation function i relativizes first order quantifiers to a predicate $\psi(x)$ (i.e., if $i(\forall x\varphi)$ is $\forall x(\psi(x) \to i(\varphi))$ and $i(\exists x\varphi)$ is $\exists x(\psi(x) \wedge i(\varphi))$, then i should relativize second order quantifiers to the predicate $\forall x (Xx \to \psi(x))$ (i.e., $i(\forall X\varphi)$ should be $\forall X(\forall x(Xx \to \psi(x)) \to i(\varphi))$ and $i(\exists X\varphi)$ should be $\exists X(\forall x(Xx \to \psi(x)) \wedge i(\varphi))$. Since interpretation functions preserve basic logical form, an interpretation of Γ in Γ' will allow us to define an isomorphism between γ and a sub-structure of γ'. So my notion of structural occurrence is essentially the same as the one Resnik introduces.

112. For a criticism of this usage, see Susan C. Hale, "Spacetime and the Abstract/Concrete Distinction."

113. Resnik, "Mathematics as a Science of Patterns: Ontology and Reference," p. 540.

114. Cf. Scott, "Axiomatizing Set Theory," and Richard Montague, "Set Theory and Higher-Order Logic."

BIBLIOGRAPHY

Aleksandrov, A. D., A. N. Kolmogorov, and M. A. Lavrent'ev, eds. *Mathematics: Its Content, Methods, and Meaning.* Cambridge, Mass.: M.I.T. Press, 1969.
Angelelli, Ignacio. "Abstraction, Looking-Around and Semantics," *Studia Leibnitiana* Sonderheft 8 (1979): 108-123.
Bartle, Robert G. *The Elements of Real Analysis.* New York: John Wiley & Sons, 1976.
Bealer, George. "Foundations Without Sets," *American Philosophical Quarterly* 18 (1981): 347-353.
———. *Quality and Concept.* Oxford: Clarendon Press, 1982.
Bell, J. L. *Boolean-Valued Models and Independence Proofs in Set Theory.* Oxford: Clarendon Press, 1985.
Benacerraf, Paul. "What Numbers Could Not Be," *Philosophical Review* 74 (1965): 47-73. Reprinted in *Philosophy of Mathematics: Selected Readings,* ed. Benacerraf and Putnam, 272-294.
Benacerraf, Paul, and Hilary Putnam, eds. *Philosophy of Mathematics: Selected Readings.* Cambridge: Cambridge University Press, 1983.
Black, Max. "The Elusiveness of Sets," *Review of Metaphysics* 24 (1971): 614-636.
Bonevac, Daniel A. *Reduction in the Abstract Sciences.* Indianapolis: Hackett, 1982.
Boolos, George. "The Iterative Conception of Set," *Journal of Philosophy* 68 (1971): 215-231. Reprinted in *Philosophy of Mathematics: Selected Readings,* ed. Benacerraf and Putnam, 486-502.
———. "To Be is to Be a Value of a Variable (or to Be Some Values of Some Variables)," *Journal of Philosophy* 81 (1984): 430-449.
———. "Nominalist Platonism," *Philosophical Review* 94 (1985): 327-344.
Boyer, Carl B. *The History of the Calculus and its Conceptual Development.* New York: Dover, 1959.
Burali-Forti, Cesare. "A Question on Transfinite Numbers." In *From Frege to Gödel: A Sourcebook in Mathematical Logic, 1879-1931,* ed. van Heijenoort, 104-111.
Cantor, Georg. *Contributions to the Founding of the Theory of Transfinite Numbers.* New York: Dover, 1955.

———. "Letter to Dedekind." In *From Frege to Gödel: A Sourcebook in Mathematical Logic, 1879-1931*, ed. van Heijenoort, 113-117.
Cohen, Paul J. *Set Theory and the Continuum Hypothesis*. Reading, Mass.: Benjamin-Cummings, 1966.
———. "Comments on the Foundations of Set Theory." In *Axiomatic Set Theory*, ed. Scott, 9-15.
Dauben, Joseph W. "The Trigonometric Background to Georg Cantor's Theory of Sets," *Archive for the History of the Exact Sciences* 7 (1971): 181-216.
Davis, Philip J., and Reuben Hersh. *The Mathematical Experience*. Boston: Birkhäuser, 1981.
Dedekind, Richard. *Essays on the Theory of Numbers*. New York: Dover, 1963.
Detlefsen, Michael. *Hilbert's Program*. Dordrecht: D. Reidel, 1986.
Dummett, Michael. "The Philosophical Basis of Intuitionistic Logic." In *Logic Colloquium '73*, ed. H. E. Rose and J. C. Shepherdson, 5-40. Amsterdam: North-Holland, 1975. Reprinted in *Philosophy of Mathematics: Selected Readings*, ed. Benacerraf and Putnam, 97-129; and in Dummett, *Truth and Other Enigmas*, 215-247.
———. *Elements of Intuitionism*. Oxford: Clarendon Press, 1977.
———. *Truth and Other Enigmas*. Cambridge, Mass.: Harvard University Press, 1978.
———. *The Interpretation of Frege's Philosophy*. Cambridge, Mass.: Harvard University Press, 1981.
Edwards, Charles H. *The Historical Development of the Calculus*. New York: Springer-Verlag, 1979.
Fitch, Frederic B. "Attribute and Class." In *Philosophic Thought in France and the United States*, ed. M. Farber. Albany, New York: SUNY Press, 1968.
Fraenkel, Abraham A., Yehoshua Bar-Hillel, and Azriel Levy. *Foundations of Set Theory*. Amsterdam: North-Holland, 1973.
Frege, Gottlob. *The Foundations of Arithmetic*. Evanston, Ill.: Northwestern University Press, 1980.
———. "Review of Dr. E. Husserl's *Philosophy of Arithmetic*." In *Husserl: Expositions and Appraisals*, ed. Frederick Elliston and Peter McCormick, 314-324. Notre Dame, Ind.: University of Notre Dame Press, 1977.
Gödel, Kurt. *The Consistency of the Continuum Hypothesis*. Princeton, N.J.: Princeton University Press, 1940.
———. "What is Cantor's Continuum Problem?" In *Philosophy of Mathematics: Selected Readings*, ed. Benacerraf and Putnam, 470-485.
Goodman, Nicolas D. "A Genuinely Intensional Set Theory." In *Intensional Mathematics*, ed. S. Shapiro, 63-79. Amsterdam: North-Holland, 1985.
Greene, William Chase, ed. *Scholia Platonica*. Haverford, Penn.: American Philological Association, 1938.
Hale, Susan C. "Spacetime and the Abstract/Concrete Distinction," *Philosophical Studies* 53 (1988): 85-102.
Hallett, Michael. *Cantorian Set Theory and Limitation of Size*. Oxford: Clarendon Press, 1984.

Hiller, Edward, ed. *Theonis Smyrnaei: Expositio rerum mathematicarum ad legendum Platonem utilium.* Leipzig: Teubner, 1878.
Husserl, Edmund. *Logical Investigations.* Atlantic Highlands, N.J.: Humanities Press, 1970.
———. *Formal and Transcendental Logic.* The Hague: Martinus Nijhoff, 1978.
Kitcher, Philip. "The Plight of the Platonist," *Noûs* 12 (1978): 119–136.
———. *The Nature of Mathematical Knowledge.* New York: Oxford University Press, 1984.
Klein, Jacob. *Greek Mathematical Thought and the Origin of Algebra.* Cambridge, Mass.: M.I.T. Press, 1968.
Kreisel, Georg. "Two Notes on the Foundations of Set Theory," *Dialectica* 23 (1969): 93–114.
———. "Observations on Popular Discussions of Foundations." In *Axiomatic Set Theory,* ed. Scott, 189–198.
Kreisel, Georg, and J. L. Krivine. *Elements of Mathematical Logic: Model Theory.* Amsterdam: North-Holland, 1971.
Kunen, Kenneth. *Set Theory: An Introduction to Independence Proofs.* Amsterdam: North-Holland, 1980.
Kuratowski, Kazimierz. "Sur la notion d'ordre dans la théorie des ensembles," *Fundamenta mathematicae* 2 (1921): 161–171.
Lakatos, Imre, ed. *Problems in the Philosophy of Mathematics.* Amsterdam: North-Holland, 1967.
———. *Proofs and Refutations: The Logic of Mathematical Discovery.* Cambridge: Cambridge University Press, 1976.
Lear, Jonathan. "Sets and Semantics," *Journal of Philosophy* 74 (1977): 86–102.
Loemker, Leroy E., ed. *Gottfried Wilhelm Leibniz: Philosophical Papers and Letters.* Dordrecht: D. Reidel, 1969.
Lorenzen, Paul. "Equality and Abstraction," *Ratio* 4 (1962): 85–90.
———. *Formal Logic.* Dordrecht: D. Reidel, 1965.
Maddy, Penelope. "Believing the Axioms," *Journal of Symbolic Logic* 53 (1988): 481–511, 736–764.
Marcus, Ruth Barcan. "Classes, Collections, and Individuals," *American Philosophical Quarterly* 11 (1974): 227–232.
Mayberry, John. "What Are Numbers?" *Philosophical Studies* 54 (1988): 317–354.
Mirimanoff, Dimitry. "Les antinomies de Russell et de Burali-Forti et le problème fondamental de la théorie des ensembles," *L'Enseignement mathématique* 19 (1917): 37–52.
Montague, Richard. "Set Theory and Higher-Order Logic." In *Formal Systems and Recursive Functions,* ed. J. N. Crossley and M. A. E. Dummett, 131–148. Amsterdam: North-Holland, 1965.
Pareigis, Bodo. *Categories and Functors.* New York: Academic Press, 1970.
Parsons, Charles. "What is the Iterative Conception of Set?" In *Logic, Foundations of Mathematics, and Computability Theory,* ed. R. E. Butts and

J. Hintikka. Dordrecht: D. Reidel, 1977. Reprinted in *Philosophy of Mathematics: Selected Readings*, ed. Benacerraf and Putnam, 503-529.

———. "The Structuralist View of Mathematical Objects," *Synthese* (forthcoming).

Peano, Giuseppe. "Le definizioni per astrazione," *Mathesis Società Italiana di Matematica, Bollettino* 7 (1915): 106-120.

Poincaré, Henri. *Science and Hypothesis*. New York: Dover, 1952.

Pollard, Stephen. "A Peculiarity of the Empty Set," *Southern Journal of Philosophy* 23 (1985): 355-360.

———. "Plural Quantification and the Iterative Concept of Set," *Philosophy Research Archives* 11 (1986): 579-587.

———. "Transfinite Recursion in a Theory of Properties," *Zeitschrift für mathematische Logik und Grundlagen der Mathematik* 32 (1986): 307-314.

———. "Identity Criteria," *Logique et Analyse* 29 (1986): 373-380.

———. "What is Abstraction?" *Noûs* 21 (1987): 233-240.

———. "Weyl on Sets and Abstraction," *Philosophical Studies* 53 (1988): 131-140.

———. "Plural Quantification and the Axiom of Choice," *Philosophical Studies* 54 (1988): 393-397.

Pollard, Stephen, and Norman M. Martin. "Mathematics for Property Theorists," *Philosophical Studies* 49 (1986): 177-186.

Rang, Bernhard, and Wolfgang Thomas. "Zermelo's Discovery of the 'Russell Paradox'," *Historia Mathematica* 8 (1981): 15-22.

Resnik, Michael D. "Mathematical Knowledge and Pattern Cognition," *Canadian Journal of Philosophy* 5 (1975): 25-39.

———. "Mathematics as a Science of Patterns: Ontology and Reference," *Noûs* 15 (1981): 529-550.

———. "Mathematics as a Science of Patterns: Epistemology," *Noûs* 16 (1982): 95-105.

———. "Second-Order Logic Still Wild," *Journal of Philosophy* 75 (1988): 75-87.

———. "Mathematics from the Structural Point of View," *Revue Internationale de Philosophie* 42 (1988): 400-424.

Robinson, Abraham. "Formalism 64." In *Logic, Methodology and Philosophy of Science*, ed. Yehoshua Bar-Hillel. Amsterdam: North-Holland, 1965.

Russell, Bertrand. *The Principles of Mathematics*. 2nd ed. New York: W. W. Norton & Co., 1964.

———. *Introduction to Mathematical Philosophy*. New York: Simon and Schuster, 1971.

Schock, Rolf. "New Foundations for Concept Theory." In *Library of Theoria*, ed. S. Hallden, No. 12. Lund: CWK Gleerup, 1969.

Scott, Dana S., ed. *Axiomatic Set Theory*. Providence, R.I.: American Mathematical Society, 1971.

———. "Axiomatizing Set Theory." In *Axiomatic Set Theory*, ed. Thomas J. Jech, 207-214. Providence, R.I.: American Mathematical Society, 1974.

Shapiro, Stewart. "Second-Order Languages and Mathematical Practice," *Journal of Symbolic Logic* 50 (1985): 714-742.
Sieg, Wilfried. "Hilbert's Program Sixty Years Later," *Journal of Symbolic Logic* 53 (1988): 338-348.
Simons, Peter. "Numbers and Manifolds" and "Plural Reference and Set Theory," in *Parts and Moments: Studies in Logic and Formal Ontology*, ed. Barry Smith, 160-260. Munich: Philosophia Verlag, 1982.
Simpson, Stephen G. "Partial Realizations of Hilbert's Program," *Journal of Symbolic Logic* 53 (1988): 349-363.
Stenius, Erik. "Sets," *Synthese* 27 (1974): 161-188.
Struik, D. J., ed. *A Source Book in Mathematics, 1200-1800*. Cambridge, Mass.: Harvard University Press, 1969.
Sylla, Edith Dudley. "The Oxford Calculators." In *The Cambridge History of Later Medieval Philosophy*, ed. Norman Kretzmann, Anthony Kenny, and Jan Pinborg, 540-563. Cambridge: Cambridge University Press, 1982.
Tarski, Alfred. "A General Method in Proofs of Undecidability." In *Undecidable Theories*, Tarski, A. Mostowski, and R. Robinson. Amsterdam: North-Holland, 1953.
Tiles, Mary. *The Philosophy of Set Theory: An Historical Introduction to Cantor's Paradise*. Oxford: Basil Blackwell, 1989.
Ulam, Stanislaw M. *Adventures of a Mathematician*. New York: Charles Scribner's Sons, 1976.
Van Aken, James. "Axioms for the Set-Theoretic Hierarchy," *Journal of Symbolic Logic* 51 (1986): 992-1004.
van Heijenoort, Jean, ed. *From Frege to Gödel: A Sourcebook in Mathematical Logic, 1879-1931*. Cambridge, Mass.: Harvard University Press, 1967.
Wagner, Steven J. "The Rationalist Conception of Logic," *Notre Dame Journal of Formal Logic* 28 (1987): 3-35.
Wang, Hao. "The Concept of Set." In *From Mathematics to Philosophy*. London: Routledge and Kegan Paul, 1974. Reprinted in Benacerraf and Putnam, 530-570.
Weyl, Hermann. *The Continuum*. Kirksville, Mo.: Thomas Jefferson University Press, 1987.
―――. *Philosophy of Mathematics and Natural Science*. Princeton, N.J.: Princeton University Press, 1949.
Youschkevitch, A. P. "The Concept of Function up to the Middle of the 19th Century," *Archive for the History of the Exact Sciences* 16 (1976): 37-85.
Zermelo, Ernst. "Investigations in the Foundations of Set Theory I." In *From Frege to Gödel: A Sourcebook in Mathematical Logic, 1879-1931*, ed. van Heijenoort, 199-215.

INDEX OF NAMES

Aleksandrov, A. D., 165
Angelelli, Ignacio, 168
Archimedes, 15
Aristotle, 16-17, 36
Ayer, A. J., xi

Bartle, Robert G., 165
Bealer, George, 50, 54, 167
Bell, J. L., 168
Benacerraf, Paul, 143, 146-147, 169, 170
Bernoulli, David, 27
Bernoulli, Jean, 25, 27
Black, Max, 42-47, 49, 51-52, 62, 138-140, 167
Bonevac, Daniel A., xii, 63, 168
Boolos, George, 99-100, 103-107, 110-111, 113, 120, 132, 137, 169
Boyer, Carl B., 25, 165, 166
Bradwardine, Thomas, 17-18, 20, 32
Burali-Forti, Cesare, 87, 169, 170
Burton, Patricia, xii

Cantor, Georg, 7, 9-10, 27-33, 38, 67-68, 80, 87, 111, 166, 169
Cohen, Paul J., 65-66, 69-71, 110, 165, 168, 169
Condorcet, Marquis de, 26

D'Alembert, Jean, 26
Dauben, Joseph W., 28, 30, 166
Davis, Philip J., 165, 168
Dedekind, Richard, 32-38, 147-148, 151, 166, 169, 170
Descartes, René, 23
Detlefsen, Michael, 169
Diophantus, 17, 21-22
Dirichlet, Peter Lejeune, 27-28, 30

Dumbleton, John, 18, 20, 32
Dummett, Michael, 72, 112, 167, 169, 170

Edwards, Charles H., 27, 166
Eudoxus, 15
Euler, Leonhard, 25-27

Fermat, Pierre de, 23
Fitch, Frederic B., 54, 167
Fourier, Joseph, 27-28
Fraenkel, Abraham A., 165, 168
Frege, Gottlob, 34, 36-38, 56, 62, 166, 168

Galilei, Galileo, 18
Gödel, Kurt, 66, 68-71, 78, 80, 106, 111, 166, 168
Goodman, Nicolas D., 54, 167
Graber, Robert Bates, xii
Greene, William Chase, 165

Hale, Susan C., 171
Hallett, Michael, 29, 166, 170
Hersh, Reuben, 165, 168
Heytesbury, William, 17-18, 20, 32
Hilbert, David, 80, 169
Hiller, Edward, 165
Husserl, Edmund, 62, 66, 151, 168

Jubien, Michael, 54, 167

Kant, Immanuel, 13
Kitcher, Philip, 13, 143, 165, 167, 170

Index of Names

Klein, Jacob, 17, 20, 22, 165
Kreisel, Georg, 168, 169
Krivine, Jean-Louise, 169
Kunen, Kenneth, 66, 69, 165, 168
Kuratowski, Kazimierz, 131, 170

Lagrange, Joseph Louis, 26
Lakatos, Imre, 13, 165, 170
Lear, Jonathan, 168, 169
Leibniz, Gottfried Wilhelm, 15, 23-25
Loemker, Leroy E., 24, 165
Lorenzen, Paul, 60, 168

Maddy, Penelope, 168
Marcus, Ruth Barcan, 167
Martin, Norman M., v, xii, 54, 167
Mayberry, John, 165, 167
Mirimanoff, Dimitry, 119, 126-127, 170
Montague, Richard, 164, 171

Newton, Isaac, 15, 23-25

Oresme, Nicole, 19-21, 23, 32

Pareigis, Bodo, 165
Parsons, Charles, 170, 171
Peano, Giuseppe, 60, 168
Plato, 15, 29
Poincaré, Henri, 138, 170
Putnam, Hilary, 169, 170

Quine, Willard Van Orman, 97

Rang, Bernhard, 169
Resnik, Michael D., 148-150, 169, 171
Robinson, Abraham, 70-71, 110, 168, 169

Russell, Bertrand, 28, 34, 38, 102, 138, 166, 167, 170

Schock, Rolf, 54, 167
Scott, Dana S., 137, 164, 168, 170, 171
Shapiro, Stewart, 170
Sieg, Wilfried, 169
Simons, Peter, 45, 52, 54, 100, 109, 167
Simpson, Stephen G., 169
Stenius, Erik, 138-143
Struik, D. J., 166
Swineshead, Richard, 17-18, 20, 32
Sylla, Edith Dudley, 18, 165

Tarski, Alfred, 57, 167
Theaetetus, 15
Theon of Smyrna, 15-16
Thomas, Wolfgang, 169
Tiles, Mary, 168

Ulam, Stanislaw M., 165

Van Aken, James, 137, 161, 170
van Heijenoort, Jean, 166
Viète, François, 20-25, 32
von Neumann, John, 119

Wagner, Steven J., 170
Wallis, John, 24
Wang, Hao, 170
Weierstrass, Karl, 32
Weyl, Hermann, 3, 33, 60, 165, 166, 168

Youschkevitch, A. P., 166

Zermelo, Ernst, 34, 51, 54, 87, 89, 90, 107, 118, 125, 167, 169